Instructor's Manual and Test-I[tem]

for Stephen Marshak's

Earth: Portrait of a Pla[net]

Instructor's Manual and Test-Item File

for Stephen Marshak's

Earth: Portrait of a Planet

John Werner

W. W. Norton & Company

New York   London

Copyright © 2001 by W. W. Norton & Company, Inc.

All rights reserved
Printed in the United States of America
First Edition

ISBN 0-393-97729-3 (pbk.)

W. W. Norton & Company, Inc., 500 Fifth Avenue,
　　New York, N.Y. 10110
　　　　　www.wwnorton.com
W. W. Norton & Company Ltd., Castle House, 75/76 Wells Street,
　　London W1T 3QT

1 2 3 4 5 6 7 8 9 0

Contents

Part I   **OUR ISLAND IN SPACE**
Chapter 1   Cosmology and the Birth of Earth   1
Chapter 2   Journey to the Center of the Earth   9
Chapter 3   Drifting Continents and Spreading Seas   17
Chapter 4   The Way the Earth Works: Plate Tectonics   24

Part II   **EARTH MATERIALS**
Chapter 5   Patterns in Nature: Minerals   35
Chapter 6   Up from the Inferno: Magma and Igneous Rocks   42
Chapter 7   A Surface Veneer: Sediments and Sedimentary Rocks   51
Chapter 8   Change in the Solid State: Metamorphic Rocks   63

Part III   **TECTONIC ACTIVITY OF A DYNAMIC PLANET**
Chapter 9   The Wrath of Vulcan: Volcanic Eruptions   72
Chapter 10   A Violent Pulse: Earthquakes   81
Chapter 11   Crags, Cracks, and Crumples: Crustal Deformation and Mountain Building   91

Part IV   **HISTORY BEFORE HISTORY**
Chapter 12   Deep Time: How Old Is Old?   98
Chapter 13   A Biography of Earth   109

Part V   **EARTH RESOURCES**
Chapter 14   Squeezing Power from a Stone: Energy Resources   120
Chapter 15   Riches in Rocks: Mineral Resources   130

Part VI   **PROCESSES AND PROBLEMS AT THE EARTH'S SURFACE**
Chapter 16   Unsafe Ground: Landslides and Other Mass Movements   136
Chapter 17   Streams and Floods: The Geology of Running Water   141
Chapter 18   Restless Realm: Oceans and Coasts   149
Chapter 19   A Hidden Reserve: Groundwater   158
Chapter 20   An Envelope of Gas: Earth's Atmosphere and Climate   166
Chapter 21   Dry Regions: The Geology of Deserts   178
Chapter 22   Amazing Ice: Glaciers and Ice Ages   187
Chapter 23   Global Change in the Earth System   196

Answers to Multiple-Choice Questions   203

Chapter 1
Cosmology and the Birth of Earth

Learning objectives

1. Students should be aware of the Big Bang and the major evidence supporting it. Distant galaxies are uniformly red shifted, rather than blue shifted; this implies that they are all moving away from us. The farthest galaxies are those that are most strongly red shifted, meaning that they are receding fastest. Extrapolation of velocities and trajectories into the past suggests that all matter in the Universe was contained in a single point, about 12 to 15 billion years ago (most recent estimates favor 12 billion). At that time, the Universe explosively came into existence (hence the name Big Bang); radiation from the Big Bang still can be perceived in all directions of the sky (even apparently empty space) with a radio telescope.

2. Stars, including our Sun, are nuclear fusion reactors. For most of their life histories (on the order of billions of years), hydrogen atoms are fused together to form helium. Later stages in stellar evolution include fusion of helium atoms and other, heavier elements; ultimately, iron is the heaviest element that can be produced through fusion reactions within stars.

3. After their cycles of fusion are complete, large stars violently explode, forming elements heavier than iron and leaving behind a residue of diffuse nebulae, which may be recycled to form a new star at some point in the future. These explosive events are termed novas and supernovas because some were bright enough to be seen as "new stars" in the night sky. Historically, a few supernovas have been bright enough to see during daylight.

4. Our Sun is approximately 5 billion years old and is expected to continue fusing helium as it does today for about another 5 billion years. Except for Pluto, all planetary orbits are coplanar, and all planets orbit in the same direction (counterclockwise as viewed from above Earth's north pole). These facts imply simultaneous planetary formation from a swirling nebula surrounding the Sun (the similarities in orbits would then be a natural result of conservation of angular momentum). The planets accreted from this nebula through gravitational attraction and haphazard collisions.

5. The terrestrial planets (Mercury, Venus, Earth, and Mars) are relatively small, dense, and rocky worlds because solar winds from the nearby Sun expelled most of the superabundant (but very light) elements, hydrogen and helium. The gas giant planets (Jupiter, Saturn, Uranus, Neptune) retained these elements and are thus much larger and much less dense (Saturn is less dense than water).

6. Our Moon, responsible for Earth's tides, has a chemistry similar to Earth's mantle; the Moon is thought to have originated from debris accumulated when a Mars-sized body impacted the Earth very early in Earth's history.

Summary from the text

Most Greek philosophers favored a geocentric Universe concept, which placed the Earth at the center of the Universe, with the planets and Sun orbiting around the Earth within a celestial sphere speckled with stars. The heliocentric (Sun-centered) Universe concept was not widely accepted.

Ptolemy's mathematical formulations led to the acceptance of the geocentric Universe concept throughout the Middle Ages.

The Renaissance brought a revolution in scientific thought. The idea of a spherical Earth returned, and Copernicus reintroduced the heliocentric concept, which was then proven by Galileo. Newton introduced physical laws that allowed people to understand motion in the Universe.

Eratosthenes was able to measure the size of the Earth in ancient times, but it was not until fairly recently that astronomers accurately determined the distance to the Sun, planets, and stars. Distances in the Universe are so large that they are usually measured in light-years.

The Earth is one of nine planets orbiting the Sun, and this solar system lies on the outer edge of a slowly revolving galaxy, the Milky Way, which is composed of about 300 billion stars. The Universe contains at least hundreds of billions of galaxies.

The red shift of distant galaxies, a manifestation of the Doppler effect, indicates that all distant galaxies are moving away from the Earth. This observation leads to the expanding Universe theory. Most astronomers agree that this expansion began after the Big Bang, a cataclysmic explosion that took place between 10 and 20 billion years ago.

The first atoms (hydrogen and helium) of the Universe developed about one million years after the Big Bang. These atoms formed vast gas clouds called nebulae.

According to the nebula theory of planet formation, gravity caused clumps of gas in the nebulae to coalesce into dense, revolving balls. As these balls of gas collapsed inward, they evolved into flattened disks with a bulbous center. The protostars at the centers of these disks eventually became dense and hot enough that fusion reactions began. When this happened, they became true stars, emitting heat and light.

Heavier elements form during fusion reactions in stars; the heaviest are probably made in supernova explosions. The Earth and the life forms on it contain elements that could only have been produced during the life cycle of stars. Thus, we are all made of stardust. Several generations of stars have formed and have died since the Big Bang.

Planets developed from the rings of gas and dust, the planetary nebulae, that surrounded protostars. The gas condensed into planetesimals that then clumped together to form protoplanets, and finally true planets. In our solar system, solar wind blew lighter elements outward, where they accumulated and formed the Jovian, or gas-giant, planets. The rocky and metallic balls in the inner part of the solar system did not acquire huge gas coatings; they became the terrestrial planets.

The Moon formed from debris blasted free from Earth when our planet collided with a Mars-sized planet early during the history of the solar system.

A planet the size of the Moon or larger will assume a near-spherical shape because the warm rock inside is so soft that gravity smoothes out irregularities.

Answers to review questions

1. Why do the planets appear to move with respect to the stars?

    Stars are so relatively distant that they appear fixed with respect to one another as viewed from Earth. As Earth and the other planets traverse through their orbits around the Sun, the positions of the planets vary with respect to the "fixed" celestial "sphere."

2. Contrast the geocentric and heliocentric Universe concepts.

    The geocentric concept placed the Earth at the center of the Universe, with the Sun and the other planets revolving around it. The heliocentric concept placed the Sun as the center, with Earth and the other planets revolving around it.

3. How did Galileo's observations support the heliocentric Universe concept?

    Galileo discovered four moons orbiting Jupiter, proving conclusively that at least some objects in the solar system do not orbit Earth.

4. Describe how Foucault's pendulum demonstrates that the Earth is rotating on its axis.

    Foucault set forth a heavy pendulum and observed its long-term behavior. Slowly but surely the swing path of the pendulum appeared to rotate about a vertical axis. Inertially, unless a new force is added, the pendulum will maintain its swinging plane. Foucault concluded correctly that the Earth must have rotated in order for the plane to appear to have changed in this manner.

5. How did Eratosthenes calculate the circumference of the Earth?

    He knew that when the Sun's rays were directly overhead at the town of Syene that they were seven degrees from vertical in Alexandria, a city due north of Syene. He measured the distance between the two cities, assumed the Earth to be a sphere, and calculated a circumference of about 40,000 km (the calculation is given in the text). He was extremely close to the correct answer.

6. Describe how the parallax method can be used to estimate the distance to far objects.

    Parallax relies on the apparent fixity of distant stars within our galaxy. If a nearby star is observed during the northern summer solstice, and again during the northern winter solstice, its position with respect to the "fixed background" of stars will have changed because the Earth was at opposite sides of the Sun during these two observations. Parallax is measured as a fraction of one second of arc; the closer an object is to Earth, the greater the angle of parallax that is produced in this manner. Given the relative closeness of the Earth and Sun, the parallax effect is not very strong, though it can be accurately measured for relatively close stars.

    A parsec is defined as the distance at which an object would produce a parallax angle of one second. All stars are farther than one parsec from Earth. Note that this means that they produce parallax angles that are less than 1 arcsecond.

7. You hear the main character in a cheap science-fiction movie say he will "return ten light years from now." What is wrong with his usage of the term "light year"? What are light years actually a measure of?

Light years are a measure of distance (not time), specifically the distance light travels through a vacuum in one year. There are approximately 3.26 light years in a parsec.

8. Describe how the Doppler effect works.

Sound waves (and light waves as well) emanating from an approaching source arrive at a higher frequency than they would if the object were stationary. This frequency shift arises because each successive sound wave is emanated from a closer distance than was the previous wave (see Fig. 1.9 of the text). These high frequencies are interpreted by our brains (after transmission through our ears) as a higher pitch. Once a wave source passes an observer, its sound waves have a reduced frequency, as each wave is emitted from a slightly more distant point.

9. If you hear a train whistle's pitch growing higher and higher, is it moving toward you or away from you?

The train is moving toward you and gaining speed. Stay off the tracks!

10. What does the red shift of the galaxies tell us about their motion with respect to the Earth?

All distant galaxies are moving away from our own, with the farthest galaxies moving fastest.

11. Briefly describe the steps in the formation of the Universe and the solar system.

The Universe formed from the Big Bang, an explosion of matter and space from an infinitely dense point source (singularity). It is thought that only hydrogen and helium, among the known elements, were produced in the Big Bang. Other elements are the result of stellar fusion and explosive supernovas. Our Sun and the surrounding solar system condensed from a mostly gaseous nebula, which itself contained material from previous supernovas (as evidenced by the diversity of heavier elements present in the Sun and solar system).

At the center of this spiraling nebula, most of the mass condensed to form the Sun, which graduated from protostar status when it became sufficiently massive, and thus hot enough, to fuse hydrogen. The planets arose from gravity-driven accretion and the collisions of smaller bodies termed planetesimals and protoplanets. Hydrogen and helium were blown out of the region interior to the orbit of Jupiter by solar winds, so the terrestrial planets ended up as smaller, rocky spheres of relatively high density. Farther out, the gas giant planets incorporated abundant hydrogen and helium to become much more massive, but less dense.

12. How is a supernova different from a normal star?

A supernova is the very luminous result of explosion at the end of the life cycle of a massive star. Much of the star's material forms nebulae and may become incorporated into new stars in the future.

13. Why do the inner planets consist mostly of rock and metal, but the outer planets mostly of gas?
    Gases, specifically hydrogen and helium, are more abundant in the solar system as a whole. However, solar wind ejected these gases from the inner solar system, so they could not be incorporated into the terrestrial planets. The terresetrial planets are rocky and metallic by default; these substances were sufficiently dense that they could accrete within the inner solar system.

14. Why are all the planets in the solar system (except Pluto) orbiting the Sun in the same direction and in the same plane?
    All planets coalesced from a discoid-shaped, swirling nebula surrounding the globular region which eventually became our Sun.

15. Describe how the Moon was formed.
    The Moon formed when a Mars-sized body impacted Earth early in the history of the solar system. The force of impact ejected material similar in composition to Earth's mantle. This mantle-like mass cooled and hardened, resulting in our Moon.

16. Why is the Earth round?
    Self-gravity forces objects the size of Earth to be nearly spherical (the most compact shape, minimizing the distances of points from the center).

Test bank

1. The concept of a **geocentric** Universe was first proposed during the time of the ancient Greeks. This model _____.
    A. was abandoned during the time of the Roman Empire and would never be widely held again
    B. was held to be true by thinkers throughout the Middle Ages, up until the Renaissance
    C. was rediscovered by the Polish astronomer Copernicus and has been the accepted model of the Universe ever since
    D. has been proven by NASA space photos

2. In the **heliocentric** model of the Universe, _____.
    A. Earth orbits around the Sun
    B. the Sun orbits around Earth
    C. Earth is a stationary planet
    D. Mercury and Venus orbit around the Sun, but all other planets orbit about Earth

3. In our current understanding of the Big Bang, _____.
    A. Earth is much older than the rest of the Universe
    B. the Universe is considerably older than Earth
    C. Earth and the Universe formed at about the same time
    D. there is no way of knowing how old the Universe might be

4. As the Universe has evolved, _____.
   A. hydrogen has been lost through fusion to form helium within stars
   B. hydrogen concentration has increased through the fission of helium atoms
   C. hydrogen concentration has increased through the fusion of helium atoms
   D. the number of hydrogen atoms has likely remained constant

5. From choices below, the best estimate of the age of the Universe is _____ old.
   A. five million
   B. five billion
   C. fifteen billion
   D. one hundred billion

6. The Big Bang theory states that _____.
   A. all stars will end their lives explosively as supernovas
   B. the Earth formed through a series of violent collisions
   C. meteors were responsible for the extinction of the dinosaurs
   D. all matter in the Universe was once confined to a single point

7. Strong evidence that the Universe is expanding comes from the fact that the light emitted from distant galaxies appears to be _____.
   A. red shifted
   B. blue shifted
   C. green shifted
   D. None of the above are correct.

8. Red shifts and blue shifts noted in the radiation spectra of celestial objects are analogous to the _____ effect observed in pitch as a train passes a stationary observer.
   A. Schwarzchild   B. Heidegger
   C. Doppler   D. Doppleganger

9. Since the initiation of the Big Bang, the temperature of the Universe has _____.
   A. increased   B. decreased   C. stayed about the same

10. The most abundant elements found in the Earth's rocks (such as oxygen, silicon, iron) were formed _____.
    A. by fission reactions within stars
    B. by fusion reactions within stars
    C. by direct precipitation from sea water

11. Atoms that are heavier than iron can only be produced by _____.
    A. fission reactions within stars   B. fusion reactions within stars
    C. explosion of supernovas   D. the Big Bang

12. By far the most common elements in the Universe and in our solar system are _____.
    A. nitrogen and oxygen
    B. iron and manganese
    C. hydrogen and helium
    D. hydrogen and oxygen

13. Which of the following bodies is the smallest?
    A. planets
    B. stars
    C. protoplanets
    D. planetesimals

14. Eratosthenes was the first person to accurately estimate the size of the Earth. He accomplished this feat by _____.
    A. sailing around the world and estimating his average rate of travel
    B. comparing the length of an Earth day with the distance between Earth and the Sun
    C. measuring the severity of the greatest earthquakes
    D. observing shadows simultaneously cast at two different cities that were separated by a known distance

15. The larger a planet or moon is, the more likely it is to _____.
    A. lack an atmosphere
    B. resemble a cube
    C. have a spherical shape
    D. have liquid oceans at the surface

16. Aside from Earth, the terrestrial planets are _____.
    A. Mars, Mercury, and Venus
    B. Mars, Venus, and Jupiter
    C. Jupiter, Saturn, Uranus, and Neptune
    D. Mars and Saturn

17. The gas giant, or Jovian, planets are _____.
    A. Mars, Mercury, and Venus
    B. Mars, Venus, and Jupiter
    C. Jupiter, Saturn, Uranus, and Neptune
    D. Mars and Saturn

18. The branch of science that studies the structure and history of the Universe is _____.
    A. cosmetology
    B. scientology
    C. cosmology
    D. universalism

19. As first found by Kepler, the shape of planetary orbits is most accurately described as a _____.
    A. circle
    B. oval
    C. ellipse
    D. figure eight

20. Distances to nearby stars can be accurately estimated by observing _____.
    A. how bright they appear in the night sky
    B. their angle of parallax (discrepancy of position against the background of distant stars dependent upon the point in Earth's orbit from which observations are made)
    C. how much their light has been red shifted
    D. how much hydrogen they possess

21. The circumference of the Earth is most nearly _____.
    A. 400 km   B. 4,000 km
    C. 40,000 km   D. 4,000,000 km

22. A light year is a unit that measures _____.
    A. time   B. mass
    C. distance   D. luminous intensity

23. Our Sun belongs to a galaxy known as _____.
    A. Andromeda   B. Cepheus
    C. the Milky Way   D. the Stratosphere

24. In agreement with the Big Bang theory, our Universe is _____.
    A. expanding   B. contracting   C. static (unchanging)

25. The very first stars in the history of our Universe would have differed from our Sun in that _____.
    A. they did not contain hydrogen and helium
    B. they were composed of only hydrogen and helium
    C. they would have all been smaller than our Sun
    D. even at a mass, density, and temperature comparable to our Sun, hydrogen atoms would not have fused to make helium

26. The stream of charged particles given off by the Sun, which disallowed the accumulation of hydrogen and helium in terrestrial planet formation, is called _____.
    A. the aurora borealis   B. solar wind
    C. the Sun's corona   D. the Van Allen belts

27. Chemically, the Moon is quite similar to _____.
    A. sea water   B. Earth's crust
    C. Earth's mantle   D. Earth's core

28. Foucault's experiment with a pendulum proved that _____.
    A. Earth is the center of the Universe
    B. Earth revolves around the Sun
    C. Earth rotates about an internal axis
    D. the Sun revolves around Earth

29. The only planet with an orbital plane that is not approximately parallel to that of the other planets is _____.
    A. Mars   B. Mercury   C. Uranus   D. Pluto

30. The only planet more distant from the Sun than Mars besides the gas giants is _____.
    A. Saturn   B. Neptune   C. Uranus   D. Pluto

Chapter 2
Journey to the Center of the Earth

Learning objectives

1. Students should be aware of the presence of Earth's magnetic dipole, how the magnetic field arises, and its important consequences for life on Earth.

2. Earth's nitrogen- and oxygen-dominated atmosphere starkly contrasts with those of neighboring Mars and Venus (which are mostly carbon dioxide). Earth's atmosphere is stratified, with weather and life confined to the lowest layer, the troposphere.

3. Earth is composed of a variety of materials with disparate physical properties (minerals, organics, gases, and melts). This has led to a complex physical chemistry and biochemistry, allowing both Earth's surface and its constituent life to evolve dramatically over time.

4. Earth is chemically divided into a thin, rocky crust dominated by silicate minerals, a thick mantle dominated by iron- and magnesium-rich silicates (subject locally to partial melting), and a thick, metallic core which is primarily iron (the outer portion of which is liquid). Students should know how seismic waves tell us that the outer core must be liquid.

5. Physically, the uppermost layers of Earth are the rigid lithosphere (crust and uppermost mantle) and the asthenosphere, which is softer and flows plastically. The "plates" of plate tectonic theory are discrete slabs of lithosphere, which move with respect to one another atop the weaker asthenosphere.

Summary from the text

The Earth has a magnetic field, which shields it from solar wind. Closer to Earth, the field creates the Van Allen belts, which trap cosmic rays.

A layer of gas surrounds the Earth. This atmosphere (78% nitrogen, 21% oxygen, 1% other gases) can be subdivided into distinct layers. All weather occurs in the troposphere, the layer we live in. Air pressure decreases with elevation, so 50% of the gas in the atmosphere is below 5.5 km.

The Earth consists of organic chemicals, minerals, glasses, rocks, metals, melts, and volatiles. Most rocks on Earth contain silica (SiO_2). We distinguish between silicic, intermediate, mafic, and ultramafic rocks on the basis of the proportion of silica to iron and magnesium within them. Silicic rocks have the greatest proportion of silica-rich minerals and are the least dense, while ultramafic rocks have the greatest proportion of iron- and magnesium-rich minerals and are the most dense.

The Earth's interior can be divided into three compositionally distinct layers, named in sequence from the surface down: the crust, the mantle, and the core. The first recognition of the division came from studying the density and shape of the Earth.

Pressure and temperature both increase with depth in the Earth. At the center, pressure is 3.6 million times greater than at the surface, and temperature reaches 4,300°C, nearly as hot as the surface of the Sun. The rate at which temperature increases with depth is the geothermal gradient.

Studies of seismic waves revealed the existence of sublayers in the core (outer core and inner core) and mantle (upper mantle, transition zone, and lower mantle).

The crust is a thin skin that varies in thickness from 7–10 km (beneath the ocean) to 25–70 km (beneath the continents). Oceanic crust is mafic in composition, while average continental crust is silicic to intermediate. The mantle is composed of ultramafic rock. The core consists of iron alloy.

The crust plus the uppermost part of the mantle comprise the lithosphere, a relatively rigid shell that lies over the asthenosphere, the soft layer of the mantle below.

Answers to review questions

1. Describe the shape of the magnetic field of the Earth. What does it do to the solar wind?

The typically spherical field lines associated with a magnetic dipole (such as Earth or a bar magnet) are deformed into a teardrop shape which points away from the Sun due to the influence of solar wind. Nevertheless the field effectively deflects the charged particles that make up the wind, so the vast majority do not bombard Earth.

2. What are the Van Allen radiation belts? How do they protect the Earth?

The Van Allen radiation belts are swaths of space that contain charged particles (protons, electrons, atomic nuclei) that were bombarded towards Earth either from the Sun (as solar wind) or from space (as cosmic rays). These particles are trapped within Earth's magnetic field and move along field lines, rather than hurtling towards Earth's surface. Cosmic rays and solar wind are potentially injurious to life on Earth, so the Van Allen belts are important environmental protectors.

3. With every breath you take, what are you breathing?

Mostly nitrogen and oxygen (which you use for respiration), with much smaller amounts of argon, carbon dioxide, neon, sulfur dioxide, ozone, water vapor, etc.

4. In what layer of the atmosphere does most weather occur?

The troposphere.

5. Why don't jet engines work above 25 km?

There is insufficient air pressure at that altitude; available oxygen is insufficient for burning fuel.

6. What is the proportion of land and sea on the Earth's surface?

70% of the surface is covered by the oceans, and the remaining 30% is land.

7. What is a hypsometric curve?

A hypsometric curve is a diagram depicting the proportional distribution of the Earth's solid surface over a range of altitude.

8. What is the elevation of most of Earth's surface that is above sea level? What is the elevation of most of the sea floor?

Most of the continental mass is less than 1 km above sea level. Most ocean floor is 2.5–4.5 km below sea level.

9. What are the four most abundant chemical elements in the earth?

Iron, oxygen, silicon, magnesium.

10. What are the main categories of materials that make up the Earth?

Organic chemicals, which make up the majority of living matter, are carbon- and hydrogen-based compounds (including oil and natural gas), many of which are quite complex, sometimes incorporating oxygen (as in sugars, starches, and fats), and additionally nitrogen (as in proteins), and occasionally some phosphorus and sulfur as well.

Minerals are solid, inorganic materials in which there is a fixed arrangement of atoms (this arrangement is often termed a crystalline lattice). Quartz and calcite are important, familiar examples. Mineral crystals are commonly weathered to produce fragments with rough or rounded surfaces, which are termed grains.

Rocks are cohesive aggregates of crystals or grains. Igneous rocks crystallize from molten (liquid) rock. Sedimentary rocks arise from the cementation of loose grains (sand, mud, pebbles, etc.) and through chemical precipitation (from the ocean or continental bodies of water). Metamorphic rocks arise from heat- and pressure-induced alteration of preexistent rock (without melting).

Glasses are physically solid structures in which the atoms are internally disordered (as in liquids, but without the tendency to rapidly flow). Commercial glass is produced when quartz is melted and then cooled rapidly (quenched in cool water), so that atoms cannot align themselves into the quartz crystalline arrangement before the rigidity of cooling sets in.

Metals are solids comprised of metallic elements only (to a strong approximation), such as gold, iron, and copper. (Naturally occurring metals are a subset of minerals.)

Melts are hot liquids that crystallize at surface temperatures to form igneous rocks. Melts within the Earth are termed magma; melts extruded upon the surface are termed lava.

Volatiles are substances that are stable in a gaseous state at the relatively low temperatures of Earth's surface.

11. How deep is the deepest well to ever penetrate the Earth's crust? How does this compare with the distance to the center of the Earth?

12 km, approximately three-hundredths of one percent of the distance to the Earth's center.

12. What does the average density of the Earth, and of the rocks of the crust, tell us about the density of the Earth's interior?

Crustal rocks (less than 3 g/cm^3) are substantially less dense than the Earth as a whole (~ 5.5 g/cm^3), so the interior must be considerably more dense than the crust (the density of the core is estimated to be ~ 13 g/cm^3).

13. How do we know that the Earth's mass is mostly concentrated at the center? How do we know that Earth must be largely solid?

If mass were not concentrated at the center, rotation of the Earth would cause it to deform into a more oblate (discoid) shape. If the rocky crust were floating on a liquid mantle, continental land masses would noticeably react to tides just as the oceans do.

14. What is the geothermal gradient? Why does it get hotter toward the center of the Earth?

The geothermal gradient is the rate of temperature increase with depth beneath the surface. Earth gained much of its internal heat early in its history from radioactivity, the initial collisions that formed Earth, and the differentiation that produced a dense, iron-rich core. Earth is slowly losing internal heat to the exterior and ultimately to space, through radiation at the surface, convection in the mantle, and the slow process of thermal conduction. As with food from the microwave, the exterior cools rapidly to a temperature near that of its surroundings, but the interior (especially of a massive body) is thermally stable and does not cool rapidly. Additionally, within the crust, heat continues to be added by radioactive decay.

15. Contrast continental and oceanic crust in terms of thickness and composition.

Marine crust is relatively thin (10 km or less) and primarily composed of basalt (mafic volcanic rock). Continental crust is thicker (up to 70 km) and primarily composed of intrusive igneous rocks of intermediate or silicic composition (such as granite).

16. What is the Moho, and how was it first recognized?

The Moho is the seismic definition of the crust/mantle boundary, recognized by a seismic-velocity discontinuity (surface where seismic waves abruptly propagate more rapidly upon entering the denser mantle).

17. What is the mantle composed of? How do its density and temperature change with depth?

The mantle is mostly composed of the rock peridotite, dominated by iron and magnesium-rich silicate minerals. Both temperature and density increase with depth.

18. What is the core composed of? How do we know this?

The core is an alloy dominated by iron. Study of meteorites suggests that iron, the heaviest element created by stellar fusion, is by far the most common of the dense, metallic elements in the solar system.

19. Contrast the lithosphere and asthenosphere. How are the criteria for defining lithosphere and asthenosphere different from the criteria used to recognize the crust, mantle, and core?

The lithosphere (crust and uppermost mantle) is rigid, whereas the underlying asthenosphere is ductile (less rigid). This physical distinction is in contrast to the chemical differences that mark the crust/mantle and core/mantle boundaries.

Test bank

1. Earth's surface is protected from solar wind and cosmic radiation by _____.
    A. Earth's gravitational field
    B. Earth's magnetic field
    C. a large, metallic shield launched into orbit by NASA in the 1960s
    D. a powerful stream of ions emitted by the Sun

2. The shape of Earth's magnetic field is approximately that of a _____.
    A. monopole
    B. dipole (such as that produced by a bar magnet)
    C. torus, a donut-shaped ring parallel to Earth's equator

3. Heat transfer that occurs through the movement of a fluid, driven by temperature differences among various points within the fluid, is termed _____.
    A. radiation   B. conduction
    C. convection   D. adhesion

4. Presently, Earth's atmosphere is dominated by which two gases?
    A. hydrogen and oxygen   B. carbon dioxide and methane
    C. nitrogen and oxygen   D. nitrous oxide and sulfur dioxide

5. Unlike Earth, the atmospheres of Mars and Venus are dominated by _____.
    A. chlorine   B. carbon dioxide
    C. carbon monoxide   D. methane

6. The lowermost layer of Earth's atmosphere, which sustains life and exhibits weather, is the _____.
    A. troposphere   B. mesosphere
    C. exosphere   D. ionosphere

7. In the whole Earth, the four most common elements are oxygen, silicon, magnesium, and _____.
    A. copper   B. lead   C. iron   D. zinc

8. Earth's hydrosphere consists of _____.
    A. lakes and rivers only
    B. surficial freshwater, the oceans, and groundwater
    C. a layer of hydrogen gas in the outer reaches of the atmosphere
    D. the oceans, but not rivers or lakes

9. Most continental topography lies within a range of altitude between _____.
    A. sea level and 1 km below sea level
    B. sea level and 1 km above sea level
    C. 2–5 km above sea level
    D. 3–6 km above sea level

10. Topographically, most of the ocean floor is comprised of _____.
    A. ocean trenches (5–12 km below sea level)
    B. ocean plains (2.5–4.5 km below sea level)
    C. submarine mountains (less than 2.5 km below sea level)

11. Hydrocarbons, such as petroleum and natural gas, are classified as _____.
    A. minerals   B. fluid rocks
    C. organic materials   D. alloys

12. The most common minerals within the Earth are _____.
    A. silicates   B. carbonates
    C. oxides   D. hydroxides

13. As compared to ultramafic rocks, mafic rocks have a _____.
    A. greater proportion of silica
    B. lesser proportion of silica
    C. greater proportion of iron and magnesium atoms

14. When molten material freezes so quickly that the atoms do not have sufficient time to produce an orderly arrangement, the resulting material is termed a _____.
    A. mineral   B. volatile
    C. natural glass   D. metamelt

15. A silica-rich igneous rock that has coarse crystals and makes up much of the continental crust is _____.
    A. peridotite   B. granite
    C. gabbro   D. basalt

16. As compared to the rocks that make up the crust, the Earth as a whole is _____.
    A. considerably more dense   B. considerably less dense
    C. slightly less dense   D. about the same density

17. If Earth's density were uniform throughout its interior, _____.
    A. Earth would have escaped from the Sun's orbit by now
    B. our Moon would be advancing towards Earth instead of receding from it
    C. Earth would have spun into a more oblate (discoid) shape
    D. Earth would have spun into a more prolate shape, similar to a football

18. A fracture in the crust, where rocks slide past one another, is termed a _____.
    A. fold   B. fault
    C. flying layer   D. frictional discontinuity

19. The velocities of seismic waves traveling from earthquake foci _____.
    A. are uniform throughout all layers of the Earth
    B. monotonically decrease with depth, at a consistent rate of deceleration
    C. monotonically increase with depth, at a consistent rate of acceleration
    D. generally increase with depth, occasionally making abrupt jumps termed seismic-velocity discontinuities

20. Earth's geothermal gradient is the rate of temperature change incurred by _____.
    A. increasing altitude in the atmosphere
    B. increasing depth at ocean trenches
    C. traversing from either pole towards the equator
    D. traversing down within Earth's interior

21. The boundary between the crust and mantle is marked by a seismic-velocity discontinuity called _____.
    A. the Edsel
    B. the Moho
    C. Lyell's surface
    D. the crantle

22. Ordered from top to bottom, ophiolite sequences consist of _____.
    A. clay and microfossils, basalt, and gabbro
    B. organic-rich soil, diverse sedimentary rocks, and granite
    C. basalt, clay and microfossils, and gabbro
    D. basalt, granite, and gabbro

23. Ophiolite sequences are important to geologists because they preserve _____.
    A. continental crust
    B. oceanic crust
    C. deep mantle material
    D. the asthenosphere

24. Earth's magnetic field is generated by _____.
    A. the flow of the liquid inner core
    B. B. the flow of the liquid outer core
    C. the convective flow of the mantle
    D. magnetic minerals within the crust

25. The densest layer of the Earth is the _____.
    A. crust
    B. mantle
    C. outer core
    D. inner core

26. The distinction between the crust and mantle is primarily on the basis of a difference in _____; the distinction between the lithosphere and asthenosphere is primarily on the basis of a difference in _____.
    A. chemistry (mineral content); degree of physical rigidity
    B. color; chemistry (mineral content)
    C. degree of physical rigidity; chemistry (mineral content)
    D. chemistry (mineral content); chemistry as well

27. The lithosphere is composed of the _____.
    A. crust only
    B. crust, mantle, and outer core
    C. top 100 meters of sediments and sedimentary rocks
    D. crust and the uppermost part of the mantle

28. Rock loses its rigidity when exposed to _____.
    A. the surface of the Earth
    B. temperatures greater than 140° C
    C. temperatures greater than 1,250° C
    D. temperatures less than 0° C

29. Moving into the interior of the Earth, _____.
    A. temperature and pressure both increase
    B. temperature and pressure both decrease
    C. temperature increases, but pressure stays nearly the same
    D. temperature remains remarkably constant, but pressure increases

30. Thickness of Earth's crust varies from _____.
    A. 100–500 m   B. 1–10 km   C. 5–500 km   D. 7–70 km

Chapter 3
Drifting Continents and Spreading Seas

Learning objectives

1. Students should be aware of Wegener's amassed evidence for continental drift. The fit of coastal outlines and the distributions of rocks, fossils, and ancient climatic belts all strongly suggest that the continents were once aligned to form a supercontinent named Pangaea. Wegener's ideas had few supporters during his lifetime because he could not provide a workable mechanism through which continents could move with respect to one another.

2. During the 20th century, paleomagnetic data showed that continents must have drifted, because the rocks of isolated continents produce unequal apparent polar-wander paths for the magnetic north pole. Additionally, within the rocks of a single continent, magnetic inclination angles may change through time, which is only readily explained by the continent having drifted in a northward or southward direction.

3. Continental rocks cannot plow through oceanic crust (as was suggested by Wegener). Rather, the continents are passively pushed by the activity of sea-floor spreading, in which molten rock rises at mid-ocean ridges, cools to form new oceanic crust, and spreads laterally. Simultaneously, crust is pulled downward and engulfed at deep-ocean trenches (as required by a non-expanding Earth).

4. Sea-floor spreading was proven in the late 1960s by examination of marine magnetic anomalies, which are symmetric about the mid-ocean ridges. Combined with radiometric dates, these patterns clearly show that oceanic crust is created at the ridges and spreads outward, with crustal age increasing away from the ridge axis in either direction. Areas of positive anomaly (including the ridges themselves) arise from rocks that crystallized and cooled during times when Earth's magnetic polarity was normal (the same as today's); rocks producing negative anomalies cooled during times of reversed polarity.

5. Together, the evidence for sea-floor spreading and continental drift have been combined to form the basis of our modern understanding of plate tectonics, the unifying theory of geology which explains the links between earthquakes, volcanism, and mountain building.

Summary from the text

Alfred Wegener proposed that continents had once been stuck together to form a single huge supercontinent (Pangaea) and had subsequently drifted apart. He called this idea the continental-drift hypothesis.

Wegener drew from several different sources of data to support his hypothesis: (1) coastlines on opposite sides of the ocean match up; (2) the distribution of late Paleozoic glaciation can be explained if the glaciers formed a polar ice cap over the south end of Pangaea; (3) the distribution of late Paleozoic/early Mesozoic equatorial climatic belts is compatible with the concept of Pangaea; (4) the distribution of fossil species suggests the existence of a supercontinent; (5) distinctive rock units that are now on opposite sides of the ocean formed continuous belts on Pangaea.

Despite all the observations that supported continental drift, most geologists did not initially accept the idea, because no one could explain how continents could move.

Rocks retain a record of Earth's magnetic field at the time they formed. This record is called paleomagnetism. By measuring paleomagnetism in successively older rocks, geologists found that the apparent position of Earth's magnetic pole relative to the rocks changes through time. Successive positions of the pole define an apparent polar-wander path.

Polar-wander paths are different for different continents. This observation can be readily explained by continental drift: continents move with respect to one another, while the Earth's magnetic poles remain fixed.

The invention of echo sounding permitted explorers to make detailed maps of the sea floor. These maps revealed the existence of mid-ocean ridges, deep-ocean trenches, seamount chains, and fracture zones. Heat flow is generally greater near the axis of a mid-ocean ridge.

Around 1960, Harry Hess proposed the hypothesis of sea-floor spreading. According to this hypothesis, new sea floor forms at mid-ocean ridges, above a band of upwelling mantle, then spreads symmetrically away from the ridge axis. As a consequence, ocean basins get progressively wider with time, and the continents on either side of the ocean basin drift apart. Eventually, the ocean floor sinks back into the mantle at deep-ocean trenches.

Magnetometer surveys of the sea floor revealed marine magnetic anomalies. Positive anomalies, where the magnetic field strength is greater than expected, and negative anomalies, where the magnetic field strength is less than expected, are arranged in alternating stripes.

During the 1950s, geologists documented that the Earth's magnetic field reverses polarity every now and then. The record of reversals, dated by radiometric techniques, is called the magnetic-reversal chronology.

The proof of sea-floor spreading came from the interpretation of marine magnetic anomalies. According to Vine, Mathews, and Morley, the sea floor that forms when the Earth has normal polarity creates positive anomalies, and the sea floor that forms when the Earth has reversed polarity creates negative anomalies. Anomalies are symmetric with respect to a mid-ocean ridge axis, and their widths are proportional to the duration of polarity chrons—facts that can only be explained by sea-floor spreading. Study of anomalies allows us to calculate the rate of spreading.

Drilling of the sea floor confirmed its age and served as another proof.

Answers to review questions

1. What was Wegener's continental drift hypothesis?

Wegener stated that the continents had once been contiguous, forming a supercontinent (which he termed Pangaea), and that they later moved apart to form their present configuration.

2. How does the fit of the coastlines around the Atlantic support continental drift?

The Atlantic coasts of Africa and South America are complementary; the continents clearly seem to fit together like the pieces of a jigsaw puzzle, suggesting that at one time, the two continents were linked.

3. How does the evidence of past glaciations support continental drift?

Glacial striations and glacially derived sediments suggest that India and the southern continents were linked during the late Paleozoic and were likely situated in a (southern) polar region. Further, the striations suggest that a massive ice sheet developed in Africa and radiated outward to cover this polar land mass.

4. How does the evidence of equatorial climatic belts support continental drift?

There is evidence that such environments existed in regions that are now near the north pole.

5. How does the distribution of ancient fossils support continental drift?

Many ancient terrestrial and freshwater plants and animals have widespread distributions that would be difficult to explain given the current world configuration of continental isolation, lending credence to Wegener's concept of Pangaea.

6. Why were geologists initially skeptical of Wegener's theory of continental drift?

No one foresaw how the (undeniably solid) continents could have been pushed or pulled through the (equally solid) oceanic crust.

7. Describe how the angle of inclination of the Earth's magnetic field varies with latitude. How could this information be used to determine the ancient latitude of a continent?

The magnetic field of the Earth approximates a dipole, such as a bar magnet. At the equator, the field is oriented horizontally; moving north, field lines begin to angle downward, becoming vertical over the north magnetic pole. Many ancient volcanic rocks bear magnetically susceptible, iron-rich minerals, which become aligned with Earth's magnetic field during formation, preserving information about their latitude of formation up to the present day (providing they are not sufficiently reheated to destroy this remnant magnetism).

8. How does a basalt rock become magnetized?

Iron atoms behave as tiny bar magnets, having a tendency to align with the magnetic field of the Earth. Within basaltic melt, thermal energy keeps these magnets rotating about in random orientations; Earth's field is not strong enough to overcome the wobbling and produce alignment of the atoms. Crystallization incorporates the iron atoms into mineral crystals, such as magnetite, but at crystallization temperatures, the atoms still have rotational freedom. With cooling, thermally induced random wobbling subsides, and the iron atoms align themselves with the magnetic field of the Earth. Once the rock cools below a critical temperature (sometimes called the Curie point), the atoms lose rotational freedom and preserve the orientation of magnetic field orientation at this point in time and space. This remnant magnetism is preserved until the present day unless the rock becomes reheated above the critical temperature.

9. Why did the discovery of apparent polar-wander paths show that the continents, rather than the poles, had moved?

Apparent polar-wander paths for continents that were not physically connected do not agree with one another (the pole appears to be in more than one place at a time).

10. Describe the characteristics of mid-ocean ridges, deep-ocean trenches, and seamount chains.

Mid-ocean ridges are elongate, relatively narrow chains of basaltic volcanoes that traverse the ocean floor, rising to a relief of 2–2.5 km above the surrounding abyssal plain. In some cases, a narrow axial trough runs down the center of the ridge.

Deep-ocean trenches are elongate arcs where ocean depths reach down to as much as 12 km. The trenches border chains of volcanoes at subduction zones.

Seamount chains are an elongate series of former volcanic islands that have subsided below sea level.

11. How did the observations of heat flow and seismicity support the hypothesis of sea-floor spreading?

Heat flow and seismicity are both anomalously high at mid-ocean ridges, suggesting extensive rising magma and crustal movement at these sea-floor spreading centers.

12. What is a marine magnetic anomaly? How is it detected?

A marine magnetic anomaly is the finding of a region of oceanic crust where Earth's magnetic field appears to be either slightly stronger or weaker than the global average. Anomalies are detected by a device that measures magnetic field strength (a magnetometer) and which is towed behind a ship.

13. Describe the pattern of marine magnetic anomalies across a mid-ocean ridge. How is this explained?

Over the ridge crest, Earth's magnetic field is anomalously strong. This elongate belt of positive anomaly is flanked on either side by belts of anomalously weak magnetic field strength (negative anomaly). The alternating sequence of positive and negative anomalies continues in either direction outward from the ridge, forming a pattern that possesses mirror-image symmetry about the ridge axis.

The explanation for this is that iron atoms in crystals formed in the most recent past have remnant magnetism in concert with today's global magnetic field (and are said to have normal polarity). Extra field strength derives from the alignment of all these mini-magnets reinforcing the modern dipole. The same is true for all positive anomalies, representing crystallization and cooling that took place during times when Earth's magnetic polarity was the same as it is today.

Negative anomalies are derived from bodies of rock that crystallized and cooled during times when Earth's magnetic field had a polarity opposite to today's. The iron atoms of these rocks destructively interfere with the modern dipole, weakening the observed magnetic strength.

14. How were the reversals of the Earth's magnetic field discovered? How did they corroborate the sea-floor spreading hypothesis?
Reversals were discovered when it was noted that remnant magnetism in some rocks showed orientations that would be expected if the magnetic north and south poles had switched geographic locations. The symmetric alternation of positive and negative anomalies about ridge crests, a result of polarity reversals through time, could only be logically explained through formation of oceanic crust along the ridge axis, with subsequent conveyor-belt movement at equal rates in opposite directions.

15. How did the deep-sea drilling of the Glomar Challenger confirm the sea-floor spreading hypothesis?
Sediments atop oceanic basalts become thicker away from mid-ocean ridges, and the lower most (oldest) layers become progressively older with increasing distance from the ridge as well.

Test bank

1. Wegener proposed continental drift after he observed evidence from fossils, glacial deposits, and the fit of the continents that suggested all of the continents were once _____.
   A. aligned north to south along the prime meridian during the late Cenozoic
   B. aligned east to west along the equator during the late Mesozoic through the Cenozoic
   C. combined to form a supercontinent (he termed Rodinia) in the Proterozoic
   D. combined to form a supercontinent (he termed Pangaea) in the late Paleozoic through the Mesozoic

2. Late Paleozoic glacial deposits are not found in which of the following places?
   A. India
   B. southern Africa
   C. North America
   D. South America

3. Abundant swamps led to the formation of coal in which of the following places?
   A. India
   B. southern Africa
   C. North America
   D. South America

4. Which plant genus dominated glaciated regions during the late Paleozoic and early Mesozoic?
   A. *Ginkgo*
   B. *Glossopteris*
   C. *Neuropteris*
   D. *Quercas*

5. Wegener's idea of continental drift was rejected by American geologists because _____.
   A. his English was too poor to be understood by them
   B. he could not conceive of a mechanism that would cause continents to shift positions
   C. he had relatively little evidence supporting the existence of a supercontinent
   D. the apparent fit of continental coastlines is blurred when the margins are defined by the edges of continental shelves rather than sea level

6. The difference between geographic north (direction towards the pole of the axis of rotation) and magnetic north at a given locality is termed _____.
   A. dipole   B. magnetic declination
   C. magnetic inclination D. polar wander

7. The magnetic field of Earth in the geologic past _____.
   A. is unknown, but it is assumed to have been identical to today's
   B. is known to have been constant through geologic time, due to remnant magnetization of iron-rich minerals in rocks
   C. is known to have experienced numerous polarity reversals, due to remnant magnetization of iron-rich minerals in rocks
   D. is known to have been constant through time, on the basis of theoretical calculations

8. The angle between Earth's magnetic field lines and the horizontal ground surface at a given locality is termed _____.
   A. dipole   B. magnetic declination
   C. magnetic inclination D. polar wander

9. The apparent tendency of the north (or south) magnetic pole to vary in position over time is termed _____.
   A. dipole   B. magnetic declination
   C. magnetic inclination D. polar wander

10. The apparent polar-wander paths for continents that were not connected over some span of geologic history will likely _____ concerning the positions of the ancient magnetic pole.
    A. agree   B. disagree

11. Sea-floor spreading is driven by volcanic activity _____.
    A. in the middle of abyssal plains   B. along mid-ocean ridges
    C. at the edges of continental shelves   D. along fracture zones

12. Volcanoes which have submerged beneath the surface of the sea are termed _____.
    A. mid-ocean ridges   B. seamounts
    C. fracture zones   D. continental rises

13. The thickness of clay and planktonic microskeletons is greatest _____.
    A. along mid-ocean ridges
    B. along fracture zones
    C. at the edges of ocean basins
    D. in the center of abyssal plains

14. Within the sea floor, the rate of geothermal heat flow is greatest _____.
    A. along mid-ocean ridges
    B. along fracture zones
    C. at the edges of ocean basins
    D. in the center of abyssal plains

15. Regions of the sea floor with positive magnetic anomalies were formed during times when Earth's magnetic field _____.
    A. was exceptionally strong
    B. was exceptionally weak
    C. had normal polarity
    D. had reversed polarity

16. Regions of the sea floor with negative magnetic anomalies were formed during times when Earth's magnetic field _____.
    A. was exceptionally strong
    B. was exceptionally weak
    C. had normal polarity
    D. had reversed polarity

17. Spreading rates along mid-ocean ridges _____.
    A. have been remarkably constant through time
    B. have changed through time, but are the same everywhere on Earth today
    C. have changed through time, and today vary between 1 and 10 m/yr.
    D. have changed through time, and today vary between 1 and 10 cm/yr.

18. Marine magnetic anomaly belts run parallel to _____.
    A. mid-ocean ridges
    B. fracture zones
    C. continental coastlines
    D. continental shelves

19. Marine magnetic anomaly belts are widest when and where _____.
    A. continents are joined to form supercontinents
    B. sea-floor spreading rates are relatively rapid
    C. sea-floor spreading rates are relatively slow

20. The age of oceanic crust _____ with increasing distance from a mid-ocean ridge.
    A. increases
    B. decreases

Chapter 4
The Way the Earth Works: Plate Tectonics

Learning objectives

1. The "plates" of plate tectonics are discrete slabs of lithosphere (crust and rigid portion of the mantle) that move with respect to one another. They glide over a ductile layer of the mantle termed the asthenosphere. Boundaries between plates are either convergent (where plates move towards one another, with material from one of the plates subducted into the mantle), divergent (where plates are pushed apart at a mid-ocean ridge), or transform (where plates slide past one another). Relative plate motions are on the order of a few centimeters a year (a common analogy is that these rates approximate the rate of human fingernail growth).
2. Plate motion at all three boundary types triggers earthquakes. Plate boundaries are delineated by belts of high historical earthquake frequency.
3. Volcanism is associated with both convergent (island and continental volcanic arcs) and divergent (mid-ocean ridges), but not transform boundaries.
4. Only oceanic lithosphere is dense enough to be subducted at convergent boundaries. When continental lithosphere is pushed (by a ridge) and pulled (by a leading edge of subducting oceanic lithosphere) into another continent, a mountainous collision zone is formed, and the two plates involved become sutured together. Conversely, a single large plate can become rifted apart when its lithosphere is stretched, thinned, and broken apart by a new mid-ocean ridge.

Summary from the text

The lithosphere, the rigid outer layer of the Earth, is broken into discrete plates that move relative to one another. Plates consist of the crust and the uppermost (cooler) mantle. Lithosphere plates effectively float on the underlying soft asthenosphere. Continental drift and sea-floor spreading are manifestations of plate movement.

Some continental margins are plate boundaries, but many are not. A single plate can consist of continental lithosphere, oceanic lithosphere, or both.

Most plate interactions occur along plate boundaries; the interior of plates remain relatively rigid and intact. Earthquakes delineate the position of plate boundaries.

There are three types of plate boundaries—divergent, convergent, and transform—distinguished from each other by the movement the plate on one side of the boundary makes relative to the plate on the other side.

Divergent boundaries are marked by mid-ocean ridges. At divergent boundaries, sea-floor spreading takes place and forms new oceanic lithosphere.

Convergent boundaries, also called convergent margins, are marked by deep-ocean trenches and volcanic arcs. At convergent boundaries, oceanic lithosphere of the downgoing plate is subducted beneath an overriding plate. The overriding plate can consist of either continental or oceanic lithosphere. An accretionary prism forms out of sediment scraped off the downgoing plate as it subducts.

Subducted lithosphere sinks back into the mantle. Its existence can be tracked down to a depth of about 670 km by a belt of earthquakes known as the Wadati-Benioff zone.

Transform boundaries are marked by large faults at which one plate slides past another. No new plate forms and no old plate is consumed at a transform boundary.

Triple junctions are points where three plate boundaries intersect. Hot spots are places where a plume of hot mantle rock rises from just above the core-mantle boundary and causes anomalous volcanism at an isolated volcano. As a plate moves over the mantle plume, the volcano moves off the hot spot and dies, and a new volcano forms over the hot spot. As a result, hot spots spawn seamount/island chains.

A large continent can split into two smaller ones by the process of rifting. During rifting, continental lithosphere stretches and thins. If it finally breaks apart, a new mid-ocean ridge forms and sea-floor spreading begins. Not all rifts go all the way to form a new mid-ocean ridge.

Convergent plate boundaries cease to exist when a buoyant piece of lithosphere (a continent or an island arc) moves into the subduction zone. When that happens, collision occurs. The collision between two continents yields large mountain ranges.

Ridge-push force and slab-pull force drive plate motions. Plates move at rates of about 1–15 cm per year. Modern satellite measurements can detect these motions.

Plate tectonics theory provides a basis for understanding many geologic phenomena.

Answers to review questions

1. What is a scientific revolution? How is plate tectonics an example of a scientific revolution?

A scientific revolution occurs when a new, encompassing theory (or paradigm) arises to explain existing observations about some natural phenomena in a manner distinctly superior to the preexisting body of theory, which was previously widely supported within the scientific community until its shortcomings were exposed by the novel theory. Plate tectonics, over the course of the late 1960s and 1970s, provided more satisfactory explanations for the distributions of mountain belts, volcanoes, earthquakes, and ancient rocks and fossils than any preexisting ideas and changed the way scientists view the Earth. For example, modern texts emphasize the dynamic nature of the Earth and Earth systems, whereas older texts emphasized categorical description.

2. What are the characteristics of a lithosphere plate?

The lithosphere is the rocky portion of the Earth, relatively cool and rigid as compared to underlying mantle material (the ductile asthenosphere). The lithosphere is composed of the crust and the uppermost portion of the mantle.

3. How does oceanic crust differ from continental crust in thickness, composition, and density?

Oceanic crust is thinner, more mafic (largely basalt, whereas continental crust is granitic), and more dense.

4. Describe how the principle of buoyancy can be applied to continental and oceanic lithosphere.

Less dense continental lithosphere is supported (buoyantly) by surrounding, relatively dense oceanic lithosphere, just as ice cubes are buoyed in a glass of water.

5. Contrast active and passive margins.

Active margins are continental coastlines that are also plate boundaries. Passive margins are coastlines that are not plate boundaries.

6. What are the basic premises of plate tectonics?

The lithosphere is divided into discrete plates that move with respect to one another; this motion is facilitated by the contrast in rigidity between the plates and the weak, flexible asthenosphere immediately below them. Plate motion is driven by mid-ocean ridges, where new oceanic crust is created and pushed to either side, and also by subduction zones, where older oceanic crust descends into the mantle below. The continents are more passive players, mutually colliding, sliding past one another, or drifting apart depending upon the distribution of ridges and subduction zones.

7. How do we identify a plate boundary?

Plate boundaries are marked by linear or arcoid segments of relatively high earthquake frequency (earthquake belts).

8. Describe the three types of plate boundaries.

Divergent plate boundaries exist where lithosphere on either side is moving away from the boundary. At convergent plate boundaries, lithosphere to either side comes together, bringing subduction (if oceanic lithosphere is involved) or collision (of two continental plates). At transform plate boundaries, plates slide past one another.

9. How does crust form along a mid-ocean ridge?

The high heat flux at the ridge melts mantle material to form magma, which is relatively light and rises to the surface. Some of the magma crystallizes beneath the surface (as gabbro or in thin basaltic dikes), and some erupts to form volcanic lava, which flows and ultimately solidifies to form pillow basalt.

10. What happens to the mantle beneath the mid-ocean ridge?

The heat along mid-ocean ridges is so great that rigid, lithospheric mantle rocks do not exist. Here the asthenosphere directly underlies the crust.

11. Why are mid-ocean ridges high?

The great heat flow at the ridges causes the lithosphere to be relatively light along the ridge, as compared to the denser, cooler lithosphere to either side.

12. Why is subduction necessary on a nonexpanding Earth with spreading ridges?

The introduction of new crust at the mantle causes subduction to be a topological necessity. Unless the Earth expands (or its shape is dramatically altered), surface area remains constant, so any new surficial material must be balanced by a loss of surficial material through subduction, because gravity disallows the possibility of rocks breaking off at the surface and floating into space.

13. What is a Wadati-Benioff zone, and how does it help to define the location of subducting plates?

The Wadati-Benioff zone is a region of deep (down to 670 km beneath the surface) earthquakes associated with the descent of oceanic lithosphere beneath an overriding plate. The surrounding asthenosphere is not rigid enough to generate earthquakes, so this zone definitively signals the presence of a subducted lithospheric slab.

14. Describe the major features of a convergent boundary.

The boundary is marked by a deep trench where the subducting oceanic plate bends downward in opposition to the horizontal overriding plate. Sediments scrape off the subducting plate to form an accretionary prism at the edge of the overriding plate. Behind the prism, melting associated with the subducting plate produces either a volcanic continental arc or volcanic island arc.

15. Why are transform plate boundaries required on an Earth with spreading and subducting plate boundaries?

The mid-ocean ridge is segmented, with adjacent segments offset laterally and connected by fractured segments of crust. In the region between two offset segments, the direction of motion on either side of a fracture is mutually opposed (each side being dominated by the push of volcanism emanating from the ridge segment on its side of the fracture).

16. What are two examples of famous transform faults?

The text lists the Alpine Fault in New Zealand and the San Andreas Fault in California.

17. What is a triple junction?

A point at which three plate boundaries meet.

18. Explain the processes that form a hot spot.

A plume of hot lower mantle rises to the crust and partially melts, producing an isolated active volcano (flanked on one side by extinct volcanoes produced when they were over the hot spot).

19. How can hot spots be used to track the past motions of the overlying plate?

Hot spots are relatively stable points, whereas the plates that overlie them, and which bear the associated volcanoes, are moving. Over periods of millions of years, as the plate slides over the hot spot, extinct volcanoes are ferried in the direction of plate motion, while new volcanoes are formed at the hot spot.

20. How is a seamount chain produced?
    As they move away from oceanic hot spots, extinct volcanic islands subside beneath the ocean's surface and are called seamounts.

21. Describe the characteristics of a continental rift, and give examples of where this process is occurring today.
    Continental rifts appear as elongate valleys bounded on either side by faults. Volcanism occurs along the rift as asthenosphere rises to accommodate the thinning lithosphere and melts. Rifts can be found in East Africa and in the Great Basin of the western United States.

22. Describe the process of continental collision, and give examples of where this process has occurred in the past.
    Continental rock is not dense enough to subduct beneath an overriding, opposed continental plate and will thus collide, suturing together with the adjacent plate, folding the rocks in the zone of collision, and thickening the crust locally, to form a nonvolcanic mountain range.

23. Discuss the two major forces that move lithosphere plates.
    Plates are pushed by mid-ocean ridges, as elevated lithosphere directly over the ridge pushes downward on less elevated lithosphere to either side. Plates are pulled by descending slabs at subduction zones, because old oceanic lithosphere is generally more dense than the asthenosphere into which the slabs sink.

Test bank

1. The idea that the continents have maintained fixed positions throughout time _____.
    A. was accepted by scientists until the late 1960s
    B. was replaced by the theory of plate tectonics
    C. was incorporated within the theory of plate tectonics
    D. A and B are both correct; C is incorrect.

2. The theory of plate tectonics _____.
    A. incorporates continental drift but not sea-floor spreading
    B. incorporates sea-floor spreading but not continental drift
    C. incorporates and explains both sea-floor spreading and continental drift
    D. does not incorporate sea-floor spreading or continental drift

3. Unlike the lithosphere, the asthenosphere _____.
    A. is able to flow over long periods of time
    B. has a density similar to the core
    C. varies in thickness from place to place
    D. is relatively cool

4. The lithosphere of the Earth can be bent and broken, but will not flow because it _____.
    A. is too old
    B. is too dense
    C. is too cool
    D. contains radioactive elements

5. On average, continental lithosphere _____.
    A. is thicker than oceanic lithosphere
    B. contains more mafic rocks than oceanic lithosphere
    C. is denser than oceanic lithosphere
    D. contains no crustal material, consisting solely of lithified upper mantle

6. The average thickness of continental lithosphere is about _____.
    A. 30 km   B. 60 km   C. 150 km   D. 10,000 km

7. The thickness of oceanic lithosphere is _____.
    A. uniformly 100 km
    B. greatest at the geographic poles and least near the equator
    C. greatest near the mid-ocean ridges and thins out away from the ridges
    D. least near the mid-ocean ridges and thickens away from the ridges

8. Under the theory of plate tectonics, the plates themselves are _____.
    A. discrete pieces of lithosphere at the surface of the solid Earth that move with respect to one another
    B. discrete layers of lithosphere that are vertically stacked one atop the other
    C. composed only of continental rocks, which plow through the weaker oceanic rocks
    D. very thick (approximately ¼ of the Earth's radius)

9. Within the terminology of plate tectonics, an active margin is _____.
    A. synonymous with "subduction zone"
    B. a 5-mile radius surrounding an active volcano
    C. a continental coastline that coincides with a plate boundary
    D. anywhere on Earth where earthquakes are especially frequent

10. Continental coastlines that occur within the interior of a tectonic plate are called _____.
    A. internal margins   B. passive margins
    C. active margins   D. inert margins

11. Broad, sediment-covered continental shelves are found along _____.
    A. active margins   B. passive margins

12. Tectonic plates might consist of _____.
    A. continental lithosphere only
    B. oceanic lithosphere only

C. either oceanic or continental lithosphere, or a combination of both
D. either oceanic or continental lithosphere, but not both

13. Deformed (bent, stretched, or cracked) lithosphere occurs _____.
    A. randomly over the surface of the Earth
    B. primarily within the interiors of tectonic plates
    C. primarily on the margins of tectonic plates

14. Every plate boundary can be recognized by _____.
    A. the presence of active volcanoes
    B. the presence of an earthquake belt
    C. a deep chasm which can be seen from space
    D. None of the above.

15. Tectonic plates move at rates that are approximately _____.
    A. 1–5 cm every 1,000 years   B. 1–15 cm/year
    C. 1–15 m/year   D. 10–100 m/year

16. At a divergent plate boundary, two opposed plates _____.
    A. move towards one another
    B. move away from one another
    C. slide past one another

17. At a convergent plate boundary, two opposed plates _____.
    A. move towards one another
    B. move away from one another
    C. slide past one another

18. At a transform plate boundary, two opposed plates _____.
    A. move towards one another
    B. move away from one another
    C. slide past one another

19. Mid-ocean ridges are _____.
    A. convergent plate boundaries
    B. divergent plate boundaries
    C. transform plate boundaries

20. As compared to a slowly spreading mid-ocean ridge, a rapidly spreading ridge is _____.
    A. wider
    B. narrower
    C. more silicic in lava composition

21. The blob-like bodies of lithified lava emitted by mid-ocean ridge volcanoes are termed _____.
    A. globulava   B. pillow basalt
    C. rhyolitic lenses   D. cake lava

22. The youngest sea floor occurs _____.
    A. along passive margins
    B. along active margins
    C. along mid-ocean ridges
    D. randomly distributed over the entire ocean basin

23. Oceanic lithosphere thickens away from the mid-ocean ridge primarily due to _____.
    A. the addition of new crust due to hot-spot volcanism
    B. the addition of new crust due to sedimentation
    C. the addition of new lithospheric mantle as a result of cooling
    D. reasons that geologists cannot determine at present

24. As lithosphere cools to the sides of a mid-ocean ridge, it begins to _____.
    A. rise with respect to material located closer to the ridge axis
    B. sink with respect to material located closer to the ridge axis

25. Subduction zones are _____.
    A. convergent plate boundaries
    B. divergent plate boundaries
    C. transform plate boundaries

26. Summed over the entire surface of the Earth, _____.
    A. the rate of lithospheric production at ridges is greater than the rate of lithospheric consumption at subduction zones
    B. the rate of lithospheric consumption at subduction zones is greater than the rate of lithospheric production at ridges
    C. rates of lithospheric production and consumption are equal

27. As compared to the density of the asthenosphere, the oceanic lithosphere is _____.
    A. always more dense
    B. always less dense
    C. initially more dense at the age of formation, but eventually becomes less dense
    D. initially less dense at the age of formation, but eventually becomes more dense

28. At a subduction zone, the overriding plate _____.
    A. is always composed of continental lithosphere
    B. is always composed of oceanic lithosphere
    C. may be composed of either oceanic or continental lithosphere

29. At a subduction zone, the downgoing (subducting) plate _____.
    A. is always composed of continental lithosphere
    B. is always composed of oceanic lithosphere
    C. may be composed of either oceanic or continental lithosphere

30. The Wadati-Benioff zone is a belt of earthquakes found _____.
    A. within an otherwise stable continental interior
    B. within an overriding plate at a subduction zone
    C. within a downgoing plate at a subduction zone
    D. along mid-ocean ridges

31. The Wadati-Benioff zone extends down within the mantle to a maximum depth of _____.
    A. 30 km   B. 150 km   C. 670 km   D. 990 km

32. Subducted slabs have never been detected below the Wadati-Benioff zone.
    A. true   B. false

33. Virtually all of the sediment atop a downgoing plate becomes subducted into the mantle along with the plate.
    A. true   B. false

34. With respect to the accretionary prism of a subduction zone, the volcanic arc is located toward the interior of the_____.
    A. downgoing plate   B. overriding plate

35. Sliding motion along transform faults caused the segments of the mid-ocean ridges to become dislocated with respect to one another.
    A. true   B. false

36. At a transform plate boundary, _____.
    A. old lithosphere is consumed
    B. new lithosphere is created
    C. A and B are both correct.
    D. None of the above.

37. At transform plate boundaries _____.
    A. earthquakes are common, but volcanoes are absent
    B. volcanoes are common, but earthquakes do not occur
    C. both earthquakes and volcanoes are common

38. A triple junction is a place on Earth's surface where _____.
    A. three volcanoes form a tight, triangular cluster
    B. glacial ice, continental rocks, and the ocean can be found together
    C. the boundaries of three lithospheric plates meet at a single point
    D. the boundaries of three lithospheric plates meet to form an elongate surface

39. Seamount chains are the remnants of volcanoes formed within the interior of lithospheric plates; these volcanoes are termed _____.
    A. cinder cones   B. hot spots
    C. butte volcanoes   D. deep-hole volcanoes

40. Hot spots can occur _____.
    A. only within continental plates
    B. only within oceanic plates
    C. within either continental or oceanic plates
    D. only when the thickness of the crust is less than 10 km

41. Hot spots are caused by _____.
    A. friction due to the lithosphere sliding atop the asthenosphere
    B. unusually dense concentrations of radioactive isotopes at various points in the crust
    C. hot plumes of mantle material that rises up through cooler, denser surrounding rock
    D. factors that remain completely unknown at this time

42. A seamount is _____.
    A. any portion of the ocean floor that is topographically higher than surrounding sea floor
    B. an extinct oceanic hot-spot volcano that has not yet subsided below sea level
    C. an extinct oceanic hot-spot volcano that has subsided below sea level
    D. synonymous with the term "guyot"

43. Hawaii is an example of a _____.
    A. hot-spot volcano
    B. mid-ocean ridge volcano
    C. volcanic island arc
    D. transform margin

44. If a continental rift successfully breaks a single continent into two discrete pieces, the former rift valley becomes a _____.
    A. subduction zone
    B. mid-ocean ridge
    C. transform fault zone
    D. hot spot

45. When two bodies of continental lithosphere are pulled together at a convergent boundary, the result is _____.
    A. subduction   B. collision and mountain formation

46. Most of the pushing force driving plate motion is produced _____.
    A. at mid-ocean ridges   B. at subduction zones
    C. at collision zones   D. in the interiors of continental plates

47. Most of the pulling force driving plate motion is produced _____.
    A. at mid-ocean ridges   B. at subduction zones
    C. at collision zones   D. in the interiors of continental plates

48. In addition to the two primary sources, plate motion may also be facilitated by convection occurring within the _____.
    A. crust
    B. lithosphere
    C. asthenosphere
    D. outer core

49. Slab-pull occurs because subducting slabs are _____.
    A. less mafic, and therefore less dense, than surrounding asthenosphere
    B. cooler, and therefore more dense, than surrounding asthenosphere
    C. hotter, and therefore more dense, than surrounding asthenosphere
    D. cooler, and therefore less dense, than surrounding asthenosphere

50. The rate of motion of a lithospheric plate with respect to a stationary hot spot is termed _____.
    A. relative plate velocity
    B. absolute plate velocity

Chapter 5
Patterns in Nature: Minerals

Learning objectives

1. Students should be aware of all aspects (natural occurrence, homogeneity, solid form, fixed crystalline structure, definable chemical composition) of the definition of a mineral, so that they can easily distinguish minerals from nonmineral matter.
2. Minerals are defined on the basis of physical properties; students should know all of the most commonly used properties (streak, color, hardness, luster, specific gravity, cleavage) and how these properties are assessed. The distinctive attributes of halite (salty taste), magnetite (magnetism), and calcite (reaction to weak hydrochloric acid and birefringence [transparent crystals will produce double images]) should be known as well.
3. Minerals making up rock are formed in three ways: crystallization from a melt, solid-state diffusion through a preexisting crystal, and precipitation from water.
4. The vast majority of Earth's minerals are silicates; students should know the arrangement and composition of silica tetrahedra that form the basis of these minerals.
5. Variation in the degree of oxygen sharing among tetrahedra allows for the great diversity of crystalline form in silicates; tetrahedra may be spatially isolated or linked to form chains, sheets, or three-dimensional frameworks.
6. Many minerals have commercial uses; students should be familiar with a few examples. Additionally, some minerals are treasured as gemstones.

Summary from the text

Minerals are homogeneous, naturally occurring, solid, inorganic substances with a definable chemical composition and an internal structure characterized by an orderly arrangement of atoms, ions, or molecules in a lattice.

In the crystalline lattice of a mineral, atoms occur in a specific pattern—one of nature's finest examples of ordering.

Minerals can form by the solidification of a melt, precipitation from a water solution, or diffusion through a solid.

There are over 3,000 different types of minerals, each with a name and distinctive physical properties (color, streak, luster, hardness, specific gravity, crystal form, crystal habit, and cleavage).

The unique physical properties of a mineral reflect its chemical composition and crystal structure. By observing these physical properties, you can identify minerals.

The most convenient way for classifying minerals is to group them on the basis of chemical composition. Mineral classes include: silicates, oxides, sulfides, sulfates, halides, carbonates, and native metals.

The silicate minerals are the most common on Earth. The silicon-oxygen tetrahedron, a silicon atom surrounded by four oxygen atoms, is the fundamental building block of silicate minerals.

There are several groups of silicate minerals, distinguished from one another by how silicon-oxygen tetrahedra are linked within their lattice structures.

Gems are minerals known for their beauty and rarity. The facets on cut stones used in jewelry are made by grinding and polishing these stones with a faceting machine.

Answers to review questions

1. What is a mineral, as geologists understand the term? How is this different from the everyday usage of the word?

A mineral is a naturally occurring, homogeneous, inorganic solid with a fixed crystalline structure and a definable chemical composition (limited range of chemical variability). "Minerals" in the vernacular refers to chemical elements that are essential to human nutrition.

2. Why is glass not a mineral?

Glass is atomically disordered, having no fixed crystalline arrangement. Because there is no fixed spatial arrangement for the atoms within glass, glass fails the "fixed crystalline structure" requirement in the definition of a mineral.

3. Salt is a mineral but sugar is not. Why not? Is pepper a mineral?

Neither sugar nor pepper are minerals because both of these substances are organic (both are obtained from plants).

4. Diamond and graphite have an identical chemical composition (pure carbon), yet differ radically in physical properties. Explain in terms of their crystal lattices.

In diamond, all carbon atoms are rigidly bonded together to form a three-dimensional framework, yielding the hardest substance known. In graphite, the carbon atoms are arranged in sheets, with adjacent sheets connected by very weak bonds. The sheets are thus easily separated (graphite is left behind on paper with the application of the slight pressure of a pencil).

5. In what way does the arrangement of atoms in a mineral define a pattern? How can X-rays be used to study these patterns?

The atoms within minerals occur in an orderly geometric arrangement that does not (appreciably) vary spatially within or among crystals of a single mineral. The diffraction pattern that results when X-rays are directed at crystals provides information about atom spacing within these minerals.

6. Describe the three ways that mineral crystals form.

Crystals may solidify from a melt, freezing from hot liquid rock in the formation of igneous rock. Crystals may also form through solid-state diffusion, in which new crystals are formed from the atoms that were present in a preexisting mineral. Movement of atoms or ions to form the new structure can be driven by heat and pressure; thus solid-state diffusion is associated with metamorphic rock formation. Lastly, mineral crystals can form through precipitation out of water, producing sedimentary rocks.

7. Why do some minerals contain beautiful euhedral crystals, while others contain anhedral grains?

Euhedral crystals (those with clearly defined faces and edges) develop when crystal growth occurs in unoccupied space; crystal geometry is determined by the internal lattice structure of the elements comprising the mineral. More commonly in igneous rocks, numerous crystals form more or less simultaneously in tightly packed space. In this case, the competing crystals intertwine as they crystallize, forming irregular boundaries. Such crystals lack the clearly defined faces of exemplary euhedral crystals and are termed anhedral grains (their irregular shapes are reminiscent of those of weathered grains of sediment).

8. List the principal physical properties used to identify a mineral.

Hardness (resistance to scratching), cleavage, color, luster, crystal form and habit, streak, specific gravity.

9. How can you determine the hardness of a mineral? What is the Mohs hardness scale?

Mineral hardness is determined through scratch tests. A relatively hard mineral is able to scratch a softer mineral, but the converse statement is not true. The Mohs scale of hardness is an ordinal scale of scratch resistance, with rankings from 1 (talc) at the soft end of the scale to 10 (diamond) at the hard end.

10. How do you distinguish cleavage surfaces from crystal faces on a mineral?

Cleavage planes occur in parallel sets; crystal faces are solitary, occurring only at the surface of the crystal.

11. What is the prime characteristic that geologists use to separate minerals into classes?

Minerals are divided on the basis of chemical composition (more precisely the anion(s) present in the mineral).

12. On what basis are silicate minerals further divided into distinct groups?

Silicate minerals are subdivided on the basis of the amount of oxygen sharing among silica tetrahedra and the resultant spatial arrangement of these tetrahedra.

13. What is the relationship between the way in which silicon-oxygen tetrahedra bond in micas and the characteristic cleavage of micas (the way they split into sheets)?

In micas, each silicon-oxygen tetrahedron shares three oxygen atoms with adjacent, coplanar tetrahedra. The shared silicon-oxygen bonds within these planes are very strong, but neighboring planes of tetrahedra (sheets) are much more loosely connected by weak hydrogen bonds. Cleavage in micas thus occurs along planes with a single orientation (parallel to the sheets).

14. How do sulfate minerals differ from sulfides?

The sulfates, which most commonly form through evaporation of sea water, contain the sulfate group SO_4^{2-}. Sulfides are minerals made up of a metallic element or elements bonded to the sulfide ion, S^{2-}.

15. Look at the periodic table of the elements. Note that carbon and silicon are in the same column (meaning they both have +/− 4 oxidation states and tend to form complex structures with covalent bonds). Why then are the most common minerals silicates?

In the whole Earth, silicon is far more abundant than carbon. What carbon there is on Earth is concentrated at the surface. Much of the carbon occurs within the biosphere (organisms) in the form of complex organic compounds (which are not minerals by definition). Further, the organic remains of much past life are buried as nonmineral matter (organic peat, coal, and oil). Carbon in the form of carbon dioxide exists as a gas in Earth's atmosphere, and unlike silicon, many carbon-bearing compounds are quite soluble in water.

By contrast, silica (silicon-oxygen) tetrahedra are superabundant in the crust and mantle of the Earth. Silicate structures are highly varied on the basis of the amount of oxygen sharing among neighboring tetrahedra, and a wide variety of metallic ions may be incorporated as replacements for silicon or as fillers in the empty spaces within the lattice structure. Silicate crystals are also stable as solids at relatively high temperatures found within the Earth.

16. Why are some minerals considered gems? How do you make the facets on a gem?

Gems are minerals that are valued for their aesthetic beauty. Facets on gems are cut by a faceting machine and usually do not represent original crystal faces or cleavage planes.

Test bank

1. Minerals utilized by humans as a source of metal are termed _____.
    A. metallic minerals   B. ore minerals   C. source minerals

2. Which of the following is or are a mineral or minerals?
    A. ice within a glacier on the continent of Antarctica
    B. grains of quartz on a beach at Rio de Janeiro, Brazil
    C. industrial quality diamonds produced in a laboratory
    D. A and B are correct, but not C.

3. Minerals are homogeneous substances in that _____.
    A. they all contain calcium, as does homogenized milk
    B. once a mineral is formed, it can never be destroyed
    C. a mineral's physical properties and geometry are spatially uniform (the material behaves much the same no matter where it is observed)
    D. a mineral's chemistry and physical properties commonly change drastically from one end of a crystal to the other

4. Two distinct minerals may have the same chemical formula.
   A. true	B. false

5. A single mineral may take on multiple crystalline lattice structures.
   A. true	B. false

6. Natural glass is not considered a mineral because it _____.
   A. is not homogenous
   B. is organic
   C. does not have a fixed crystalline structure
   D. can be made synthetically as well as being a naturally occurring substance

7. Which common mineral is found in most kitchens?
   A. flour	B. sugar	C. halite	D. mustard

8. Minerals in geodes form spectacular euhedral crystals because _____.
   A. all of the elements incorporated in the crystals are in plentiful supply
   B. the crystals have abundant room to grow in their hollow surroundings
   C. minerals within geodes are always framework silicates
   D. minerals within geodes always contain iron

9. The New Age practice of surrounding one's self with crystals has a strong positive effect on _____.
   A. mental health
   B. the immune system's response to illness
   C. the prospects for world peace
   D. the bank accounts of crystal healers and rock shop owners

10. The internal ordering of mineral crystals was first detected using _____.
    A. magnetic resonance imaging
    B. X-ray diffraction
    C. a scanning electron microscope (SEM)
    D. cathodized axial tomography

11. It is rare for mineral crystals to display any sort of symmetry (invariance of pattern with respect to a transformation, such as rotation or mirror-image reflection).
    A. true	B. false

12. Diamond and graphite are both polymorphs of pure silicon.
    A. true	B. false

13. The most recently formed portion of any crystal is always found _____.
    A. deep within its interior	B. on its outer edges

14. Mineral identity is diagnosed primarily through the assessment of _____.
    A. chemical properties	B. physical properties

15. The most useful diagnostic property of minerals is their color in hand sample.
    A. true	B. false

16. For the majority of minerals, the streak color obtained when the mineral is scratched against a porcelain plate is _____.
    A. likely to be diagnostic only if the mineral is hard enough to scratch porcelain
    B. more variable than the color in hand sample among crystals
    C. less variable than the color in hand sample among crystals
    D. always dark brown or black

17. The shininess of a mineral is a helpful diagnostic property termed _____.
    A. color	B. specific gravity
    C. luster	D. streak

18. Ore minerals, such as galena and hematite, tend to be distinct in their very _____.
    A. dark coloration	B. diamond-like crystal habit
    C. great specific gravity	D. metallic luster

19. Cleavage in minerals refers to _____.
    A. a tendency to break in an irregular pattern
    B. a tendency to break along planes of weakness
    C. the sharpness of edges between crystal faces
    D. the development of distinct crystal faces

20. The most abundant minerals belong to a chemical group termed _____.
    A. silicates	B. carbonates	C. halides	D. oxides

21. When in contact with hydrochloric acid, which mineral gives off bubbles of carbon dioxide gas?
    A. quartz	B. halite	C. calcite	D. fluorite

22. The silica tetrahedron that forms the backbone of all the silicate minerals is composed of silicon and what other element?
    A. magnesium	B. oxygen	C. iron	D. carbon

23. All minerals are chemical compounds (composed of more than one element).
    A. true	B. false

24. In silicate minerals, tetrahedra may be coordinated to form _____.
    A. long one-dimensional chains
    B. extensive two-dimensional sheets
    C. massive three-dimensional frameworks
    D. All of the above are possible.

25. In which type of silicate are the greatest proportion of oxygen atoms shared by pairs of adjacent tetrahedra?
    A. chain silicates
    B. framework silicates
    C. sheet silicates
    D. Sharing of oxygen atoms does not occur in silicates.

26. Minerals prized for their scarcity and beauty are termed _____.
    A. gems   B. gemstones   C. lodestones   D. firestones

27. Minerals which do not possess cleavage are said to possess _____.
    A. invulnerability   B. fracture
    C. solidity   D. massiveness

28. Gemstones are commonly found in pegmatites, which are igneous rocks that are _____.
    A. exceptionally mafic
    B. extrusive, forming from lava
    C. exceptionally coarse-grained
    D. exceptionally fine-grained

29. The facets on cut gemstones most commonly represent _____.
    A. natural cleavage planes of the mineral
    B. original crystal faces
    C. neither cleavage planes nor crystal faces, but artificially cut planar surfaces
    D. None of the above.

30. Synthetically made glass and natural quartz crystals both exhibit a fracture pattern termed _____.
    A. glassy   B. conchoidal   C. serpentine   D. obtuse

Chapter 6
Up from the Inferno: Magma and Igneous Rocks

Learning objectives

1. Students should know that igneous rocks make up a majority of the crust; basalt makes up most oceanic crust; and granite and granodiorite make up most continental crust. The ultramafic and intrusive igneous rock peridotite makes up the mantle, including the rigid, subcrustal lithosphere.
2. Igneous rocks solidify (freeze) from melt (either lava or magma); melts are derived from pressure release, the addition of volatiles, or the assimilation of heat from hot surrounding rocks or melts.
3. Igneous rocks are classified on the basis of grain size (phaneritic, aphanitic, glassy) and mineral composition (silicic, intermediate, mafic, ultramafic). Grain size and composition should be known for the standard six (granite, diorite, gabbro, rhyolite, andesite, basalt) plus peridotite.
4. Students should know the basic information summed by Bowen's reaction series, such as the high crystallization temperatures of olivine and calcium-plagioclase as compared to quartz and orthoclase.
5. Viscosity of melts varies substantially; mafic composition, high volatile content, and high temperature all tend to reduce viscosity.
6. Students should be familiar with the distinction among dikes, sills, and plutons among intrusive bodies.
7. They should be familiar with the major geologic settings for volcanism: mid-ocean ridges, island and continental volcanic arcs, hot spots, and continental rifts.

Summary from the text

Magma is liquid rock (melt) under the Earth's surface. Lava is melt that has erupted from a volcano at the Earth's surface.

Magma forms when hot rock in the Earth melts. This process only occurs under certain circumstances—where the pressure decreases (decompression), where volatiles (such as water or carbon dioxide) are added to hot rock, and where the temperature rises because heat is transferred from magma rising from the mantle into the crust.

Magma comes in a range of compositions: silicic, intermediate, mafic, and ultramafic. Mafic magma is hotter than silicic magma. The composition of magma is determined in part by the original composition of the rock from which the magma formed and by the fact that magma forms by the partial melting of rock. Contamination (the addition of material to the magma from the surroundings) and fractional crystallization (the settling out of early-formed crystals) may change the composition of a magma once it has formed.

During partial melting, only part of the source rock melts to create magma. Magma tends to be more silicic than the rock from which it was extracted, because silica-rich minerals tend to melt first.

Magma rises from the depth because of its buoyancy and because the pressure caused by the weight of overlying rock squeezes magma upward.

Magma viscosity (its resistance to flow) depends on its composition. Silicic magma is more viscous than mafic magma.

Geologists distinguish between two types of igneous rocks. Extrusive igneous rocks form from lava that erupts out of a volcano and freezes in contact with air or the ocean. Intrusive igneous rocks develop from magma that freezes inside the Earth.

Lava creates extrusive igneous rocks in a variety of ways. It may solidify to form flows or domes, or it may be exploded into the air to form ash.

Intrusive igneous rocks form when magma injects into preexisting rock (country rock) below Earth's surface. They can be tabular or sheet-like in shape, or blob shaped (plutons). Vertical sheet-like intrusions that cut across layering in country rock are dikes, and horizontal sheet-like intrusions that form parallel to layering in country rock are sills. Huge intrusions, made up of many plutons, are known as batholiths.

The rate at which intrusive magma cools depends on the depth at which it intrudes, the size and shape of the magma body, and whether there is circulating groundwater present. The cooling rate is reflected in the grain size of an igneous rock. Instantly cooled melt produces glass, less quickly cooled melt produces fine-grained rock, and slowly cooled melt produces coarse-grained rock.

Crystalline (nonglassy) igneous rocks are classified according to texture and composition. (For example, granite and rhyolite are both silicic rocks, but granite is coarse grained, while rhyolite is fine grained.) Glassy igneous rocks are classified on the basis of texture (a solid mass is obsidian, while ash that has cemented or welded together is a tuff).

The origin of igneous rocks can readily be understood in the context of plate tectonics. Magma forms at continental or island volcanic arcs along convergent margins, mostly above the subducting slab. Igneous rocks form at hot spots owing to the decompression melting of a rising mantle plume. (Where mantle-derived mafic magma rises into continental crust, it can partially melt the crust and generate silicic magma.) Igneous rocks form at rifts as a result of decompression melting of the asthenosphere below the thinning lithosphere. Igneous rocks form along mid-ocean ridges because of decompression melting of the rising asthenosphere.

Answers to review questions

1. How is the process of freezing magma similar to that of freezing water? How is it different?

Crystallization from magma is similar to freezing ice cubes in that crystals form from the liquid state upon passage of the liquid through a critical temperature (or range of temperatures). The most striking difference is in the temperatures of crystallization. Additionally, multiple minerals crystallize from magma; more mafic minerals crystallize first, and more silicic minerals crystallize later at cooler temperatures.

2. How did the first igneous rocks on the planet form?

Earth was initially so hot that its surface was molten; heat was attained from gravitational compression, radioactive isotopes, meteoric bombardment, and differentiation of the Earth's metallic core. After sufficient time, the magma surface cooled to produce the Earth's first (igneous) rocks.

3. How did Hutton show that igneous rocks came from a melt and not from water?

He noted that igneous rocks cut through sedimentary strata and occasionally deform these strata (both those that lie above and below the intrusive igneous body).

4. Describe the three processes that are responsible for the formation of magmas.

A. A decrease in pressure may cause the liquid state of a mineral assemblage to be thermodynamically favored over solid crystals.

B. The addition of volatiles (such as water and carbon dioxide) to a system of crystals (rock) may cause the liquid state to be thermodynamically favored.

C. If a system of crystals obtains heat from a nearby source (such as a neighboring body of magma), it may melt.

5. Why are there so many different types of magmas?

Magmas differ primarily with respect to chemical composition. Numerous processes can alter magmatic chemistry, and, thus, the minerals and mineral abundances of the crystals that ultimately freeze out of the melt. Changes in chemistry can occur through fractional crystallization of melt, partial melting of rock, and contamination through the melting and assimilation of surrounding host rock. In addition to these processes that produce chemical change, original (melted) source rocks vary. Melts also vary with respect to volatile content.

6. Why do magmas rise to the surface?

Magmas rise because they are less dense than the rocks that surround them, and at depth there is great pressure that squeezes magma upward, where the pressure is less.

7. What factors control the viscosity of a melt?

Silicic melts are more viscous than mafic melts, because silica tetrahedra mutually link to form complex structures that cannot move as rapidly as the simple structures found in more mafic melts. High volatile content (such as water content) lessens viscosity. Cooler magmas are more viscous than hotter magmas of the same composition.

8. What two criteria would you use to determine if an igneous rock intruded while hot, or simply eroded and became covered by sedimentary rock?

If eroded, an erosional surface should be apparent on top of the igneous body. If the igneous rock intruded into the sedimentary rock, a small rind of the sedimentary rock immediately surrounding the intrusion should have been altered by the heat of formation to produce a baked contact. Further, if a dike or pluton cuts across any of the bedding planes within the sedimentary rock, it must have formed through intrusion.

9. What factors control the rate of cooling of a magma within the crust?

Magmas that cool deep within the Earth cool more slowly than those that cool near the surface. Large, globular bodies of magma cool more slowly than those that are smaller and sheet-like. Cool groundwater can absorb heat from magma and transport the heat away from the source, greatly increasing the cooling rate.

10. How does grain size reflect the rate of cooling of a magma?

Large, thermodynamically stable crystals form when a melt cools and crystallizes gradually. Rapidly cooling and crystallizing melts do not leave sufficient time for large crystals to become organized.

11. What does the mixture of grain sizes in a porphyritic igneous rock indicate about its cooling history?

Porphyritic igneous rocks occur when initially gradual cooling within the Earth gives way to much more rapid cooling subsequent to eruption. The coarse phenocrysts are likely to be composed of calcium-rich plagioclase, which crystallizes at relatively high temperatures. The finer crystals forming the matrix are composed of minerals that crystallize at lower temperatures.

12. How does pumice differ from scoria?

Pumice, unlike scoria, is volumetrically > 50% air-filled pores.

13. Describe the way magmas are produced in subduction zones.

Below 150 km, heat causes volatiles within the subducting oceanic, lithospheric slab to be released into the ultramafic asthenosphere of the overriding plate. Partial melting of the asthenosphere produces basaltic magma, which migrates upward through the lithosphere of the overriding plate to erupt at the surface. Dependent upon the quantity of fractional crystallization that occurs during transit, the originally basaltic (mafic) magma may extrude as an intermediate or rhyolitic (silicic) melt.

14. How are hot-spot volcanoes created?

An isolated, cylindrical plume of hot material from the lower mantle rises upward. When it reaches the lithosphere, pressure is low enough to initiate partial melting, forming basaltic (mafic) melt from the ultramafic rock. The melt rises further to reside in a magma chamber in the crust until it erupts, forming a volcano.

15. Describe how magmas are produced at continental rifts.

Rifting force thins out the lithosphere locally, decreasing the pressure on the underlying asthenosphere. This pressure release triggers partial melting to yield basaltic magma, which may extrude to the surface itself or melt the surrounding crust to produce a rhyolitic melt.

16. Why does melting take place beneath the axis of a mid-ocean ridge?

   Hot asthenosphere rises to take the place of material that has spread to either side of the axis. Closer to the surface, pressure is less than at the depth from which the asthenospheric material rose, so the asthenosphere partially melts.

Test bank

1. The difference between lava and magma is that _____.
    A. magma is light in color and lava is dark
    B. magma usually has mafic composition and lava usually has silicic composition
    C. magma is found beneath the Earth's surface, whereas lava has reached the surface
    D. magma flows more quickly than lava

2. A blob-like igneous rock body that has cooled beneath the surface of the Earth is a(n) _____.
    A. guyot   B. pluton   C. lava flow   D. andesite

3. A dike is _____.
    A. a horizontal tabular intrusion that lies parallel to surrounding layers of sedimentary rock
    B. a cooled layer of lava
    C. an intrusion formed within the magma chamber of a volcano
    D. a vertical tabular intrusion that cuts across preexisting layers

4. A sill is _____.
    A. a horizontal tabular intrusion that lies parallel to surrounding layers of sedimentary rock
    B. a cooled layer of lava
    C. an intrusion formed within the magma chamber of a volcano
    D. a vertical tabular intrusion that cuts across preexisting layers

5. A volcanic neck, such as that seen at Shiprock, New Mexico, is _____.
    A. a horizontal tabular intrusion that lies parallel to surrounding layers of sedimentary rock
    B. a cooled layer of lava
    C. an intrusion formed within the magma chamber of a volcano
    D. a vertical tabular intrusion that cuts across preexisting layers

6. Geologically, melts are equivalent to both _____.
    A. silicic and mafic magmas
    B. silicic and mafic lavas
    C. lavas and magmas
    D. fine-grained and coarse-grained igneous rocks

7. Igneous rocks _____.
   A. are formed through the freezing of melt
   B. can be produced at the surface of the Earth as well as deep below the surface
   C. are the most common type of rocks within the Earth
   D. All of the above are correct.

8. Very early in Earth's history, it was so hot that the surface was entirely molten.
   A. true   B. false

9. Radioactive isotopes, differentiation of Earth's iron core, gravity-driven compression, and meteoric bombardment all caused early Earth to _____.
   A. glow brighter than the Sun
   B. be much cooler than at present
   C. be much hotter than at present
   D. be much more oblong than at present

10. Bombs, ash, and cinders are all examples of _____.
    A. intrusive igneous rocks   B. hot spots
    C. volatiles   D. pyroclastic debris

11. When magma crystallizes, _____ are formed.
    A. intrusive igneous rocks   B. extrusive igneous rocks
    C. volatiles   D. pyroclastic debris

12. The geotherm is the rate of change of _____.
    A. pressure with depth in Earth's interior
    B. temperature with depth in Earth's interior
    C. temperature with altitude in Earth's atmosphere
    D. temperature with latitude on Earth's surface

13. The formation of magma within the Earth is **not** caused by which of the following processes?
    A. decompression (drop in pressure)
    B. addition of volatiles
    C. transfer of heat from adjacent magma or very hot rocks
    D. compression (increase in pressure)

14. If a body of igneous (source) rock is subjected to partial melting, the magma that is produced is expected to be _____.
    A. identical in chemical composition to the source rock
    B. more mafic than the source rock
    C. more silicic than the source rock

15. If a body of magma is subjected to fractional crystallization, the rock that results is expected to be _____.
    A. identical in chemical composition to the magma
    B. more mafic than the magma
    C. more silicic than the magma

16. Volatiles refer to substances that _____.
    A. crystallize most rapidly out of a melt
    B. melt immediately on contact with a hot body of magma
    C. have a tendency to evaporate and are stable as gases

17. If the volatile content of magma is increased, its viscosity will _____.
    A. increase   B. decrease   C. stay the same

18. If a body of magma becomes more silicic, its viscosity will _____.
    A. increase   B. decrease   C. stay the same

19. If a body of magma cools, its viscosity will _____.
    A. increase   B. decrease   C. stay the same

20. Typically, as magma ascends through the crust, its viscosity is expected to _____.
    A. increase   B. decrease   C. stay the same

21. A volcano emits ash, which falls from the sky, settles in layers, and is eventually cemented. The resultant rock is termed _____.
    A. pumice   B. granite   C. tuff   D. ignimbrite

22. A sill that domes upward is termed a _____.
    A. dike   B. lopolith   C. laccolith   D. batholith

23. The grain size of an igneous rock is determined primarily by _____.
    A. its mineral composition
    B. its volatile content
    C. its pressure during formation
    D. how rapidly it cooled during crystallization

24. The distinction between a pluton and a batholith is _____.
    A. plutons are intrusive bodies, whereas batholiths are extrusive
    B. plutons are blob-shaped, whereas batholiths are tabular
    C. numerous batholiths make up a pluton
    D. numerous plutons make up a batholith

25. All other factors being equal, intrusive rocks that form deep within the Earth _____ than intrusive rocks that cool near the surface.
    A. are more silicic   B. contain a smaller proportion of volatiles
    C. cool more slowly   D. cool more rapidly

26. Pegmatites, which occur in dikes, are unusual among shallow intrusive rocks in that they _____.
    A. possess porphyritic texture
    B. possess exceptionally coarse grains
    C. are mineralogically identical to the extrusive rock basalt
    D. are glassy, cooling so rapidly that crystals do not have time to form

27. Obsidian _____.
    A. is volcanic glass
    B. possesses conchoidal fracture
    C. typically is silicic in composition
    D. All of the above.

28. Ash and lava fragments that cascade down the sides of a volcano eventually settle and lithify to form _____.
    A. basalt   B. andesite
    C. welded tuff   D. ash-fall tuff

29. Volcanoes that do not occur along either present or emergent plate boundaries are associated with _____.
    A. continental rifts   B. mantle hot spots
    C. mid-ocean ridges   D. subduction zones

30. A famous example of hot-spot volcanism occurs at _____.
    A. the Aleutian Islands of Alaska
    B. the Andes Mountains
    C. Hawaii
    D. Mt. St. Helens, Washington

31. An island volcanic arc occurs at _____.
    A. the Aleutian Islands of Alaska
    B. the Andes Mountains
    C. Hawaii
    D. Mt. St. Helens, Washington

32. A continental volcanic arc occurs at _____.
    A. the Aleutian Islands of Alaska
    B. the Andes Mountains
    C. Hawaii
    D. Japan

33. Coarse-grained granite is most similar in mineral composition to fine-grained _____.
    A. andesite   B. basalt   C. komatiite   D. rhyolite

34. Coarse-grained diorite is most similar in mineral composition to fine-grained _____.
    A. andesite   B. basalt   C. komatiite   D. rhyolite

35. Coarse-grained gabbro is most similar in mineral composition to fine-grained _____.
   A. andesite   B. basalt   C. komatiite   D. rhyolite

36. Most commonly, silicic igneous rocks _____.
   A. contain more iron and magnesium than intermediate rocks
   B. are lighter in color than mafic rocks
   C. are darker in color than mafic rocks
   D. are found in oceanic crust

37. Pillow basalts attain their distinctive blob-like shapes because their parent lavas do not travel far prior to solidification. This is because the parent lavas _____.
   A. are completely devoid of volatiles and thus travel slowly
   B. erupt underwater and thus cool very quickly
   C. are highly silicic and thus travel slowly
   D. are ultramafic and thus freeze at exceptionally high temperatures

38. Stoping by magmas occurs when _____.
   A. pieces of surrounding country rock are broken off and assimilated
   B. the magma stops flowing and starts to solidify
   C. the magma becomes fully solidified to form intrusive rock
   D. the magma alters a thin rind of surrounding country rock

39. Pieces of country rock that are absorbed by an intrusive body and do not melt form _____.
   A. porphyroblasts   B. phenocrysts
   C. xenoliths   D. gastroliths

40. As compared to aphanitic igneous rocks, phaneritic rocks are _____.
   A. coarser grained   B. finer grained
   C. more mafic   D. more silicic

Chapter 7
A Surface Veneer: Sediments and Sedimentary Rocks

Learning objectives

1. Students should know that, although rare in the crust as a whole, sediments and sedimentary rocks dominate solid Earth materials at the surface.

2. They should be familiar with the sedimentary portion of the rock cycle, understanding physical and chemical weathering, erosion, transport, deposition, burial, and diagenesis.

3. They should be able to distinguish among coarse-, medium- (sand), and fine-sized grains, and to evaluate sorting, sphericity, and angularity among sediments and rocks.

4. They should know what distinguishes soil from loose sediment; the typical soil profile: O, A, B, and C horizons; and how these layers are distinguished.

5. Clastic sedimentary rocks are primarily classified on the basis of grain shape. Students should be able to recognize and characterize conglomerate, breccia, sandstone, siltstone, mudstone, and shale.

6. Biologically catalyzed precipitation produces mineralized skeletons; these condense and lithify to form biochemical sedimentary rocks. Important examples are most limestones and bedded, biogenic cherts.

7. Organic sedimentary rocks (the coal series) consist of the altered remains of trees and other plants that lived in boggy environments (which allow organic matter to accumulate without too much decay).

8. Chemical sedimentary rocks arise from physically induced precipitation. As with biochemical rocks, classification is primarily on the basis of mineral content. An important subset are the evaporites, including rock salt (halite) and gypsum (small amounts of carbonate sediment can also form through evaporation). Chemical rocks can also form through the replacement of minerals during diagenesis; important replacement minerals include quartz (forming the rock called chert) and dolomite (forming dolostone).

9. The layering of rock is termed stratification (or bedding); cross beds signify deposition along a slope and can be used to indicate paleocurrent direction. Other features of interest within sedimentary rocks include fossils, ripple marks, and mud cracks.

10. The character of sedimentary rocks reflects their environments of deposition; students should be familiar with a few sedimentary environments (e.g., alluvial fan, fluvial, shallow marine, deep marine, glacial) and characteristics of the sediments deposited in these realms. Locally, environments change over time; one source of these changes is the sequence of transgressions (rising local relative sea level) and regressions (falling local relative sea level).

Summary from the text

Sediment consists of detritus (mineral grains and rock fragments derived from preexisting rock), mineral crystals that precipitate directly out of water, and shells (formed when organisms extract ions from water).

Rocks at the surface of the Earth undergo physical and chemical weathering. During physical weathering, intact rock breaks into detritus (mineral grains and rock fragments). Processes like jointing and frost wedging aid physical weathering. During chemical weathering rocks react with water and air. Chemical weathering includes reactions like dissolution, hydrolysis, and oxidation. It can produce new minerals like clay, and ions in solution.

The covering of loose rock fragments, sand, gravel, and soil at the Earth's surface is regolith. Soil differs from other types of regolith in that it has been changed by activities of organisms, by downward-percolation rainwater, and by the mixing of organic matter. Downward percolating water redistributes ions and clay.

Geologists recognize four major classes of sedimentary rocks. Clastic (detrital) rocks form from cemented-together detritus (grains or rock fragments). Biochemical rocks develop from the shells of organisms. Organic rocks consist of altered plant debris or of altered plankton remains. Chemical rocks precipitate directly from water.

Weathering, erosion, transportation, deposition, and lithification lead to the development of clastic rocks. We distinguish different types of clastic rocks on the basis of mineral content, grain size, and grain shape.

Limestone, which consists of calcite or aragonite, forms either biochemically from the shells of organisms or chemically by precipitation from water. Dolostone, which consists of dolomite, forms when water that contains magnesium reacts with limestone.

Chert forms either biochemically, from the silica shells of radiolaria (a type of plankton) or chemically, when quartz replaces other minerals in a preexisting rock. Coal (which is over 50% carbon) forms from plant detritus deposited in layers.

The precipitation of salts, when saltwater evaporates, produces deposits called evaporites.

Sedimentary rocks occur in layers called beds. The term "strata" refers to a succession of beds. Sedimentary structures include bedding (layering) of sedimentary rocks, surface features on beds formed during deposition, and the arrangement of grains in beds.

Cross beds are inclined layers formed when sediment is deposited in a current. In graded beds, which settle out of undersea avalanches called turbidity currents, the grain size of clasts progressively decreases from the base to the top of the bed. Ripple marks, mud cracks, and fossils develop on the surface of beds.

Glaciers, mountain streams and fronts, sand dunes, lakes, rivers deltas, beaches, shallow seas, and deep seas each accumulate a different sedimentary facies, a set of sedimentary rocks and structures. Thus, by studying sedimentary rocks, we can reconstruct the characteristics of past environments.

Thick piles of sedimentary rocks accumulate in sedimentary basins, regions where the lithosphere sinks, creating a depression at the Earth's surface. Basins form where the lithosphere has been loaded by a weight or where it has undergone stretching.

The sea level changes with time. Transgressions occur when the sea level rises and the coastline migrates inland. Regressions occur when the sea level falls and the coastline migrates seaward.

Answers to review questions

1. In most places, it would be impossible to empty an ocean basin. Explain the peculiar circumstances (plate tectonics, climate, oceanographic conditions, and the balance between rivers, ocean, and evaporation) that allowed the Mediterranean to dry up.

    The vast majority (< 90%) of water entering the Mediterranean comes not from freshwater rivers but rather from the saline Atlantic Ocean, through the Strait of Gibraltar. During the Pliocene epoch, approximately six million years ago, Africa advanced northward, colliding with the southern margin of Europe and closing the Strait of Gibraltar. Without the influx of ocean water, the evaporation rate in the hot, dry climate greatly exceeded the rate of riverine influx, and the level of the Mediterranean Sea progressively fell far below global mean sea level. As local sea level fell, concentrations of dissolved ions became saturated, so vast quantities of evaporate minerals (primarily halite and gypsum) precipitated out of solution.

2. How does physical weathering differ from chemical weathering?

    Physical weathering involves the mechanical breakage of rock into smaller pieces through naturally induced force and abrasion. Chemical weathering involves reactions that dissolve rock or alter its mineralogy.

3. Describe the processes that produce joints in rocks.

    When rocks in environments of great heat and pressure become exposed to environments characterized by low temperature and pressure (as would occur when overlying strata are eroded away), they deform. Being rigid, most rocks fracture along surfaces termed joints rather than becoming bent out of shape.

4. Feldspars are among the most common minerals in igneous rocks, but they are relatively rare in sediments. Why are they more susceptible to weathering? What chemical reactions are responsible for their breakdown? What common sedimentary minerals are produced from all the chemicals released by weathered feldspar?

    Feldspar crystals and grains are produced at the high temperatures associated with igneous crystallization and are more stable in such hot environments than they are at the relatively cool setting of Earth's surface. At the surface, feldspars are subjected to hydrolysis reactions, in which water is added to the mineral, producing dissolution of some cations within the feldspar and alteration of the remaining material to form clay, a group of minerals which are stable at surface conditions.

5. What kind of minerals tend to weather more quickly?

    Mafic minerals, which crystallize at very high temperatures, are highly unstable at the surface; in contrast, quartz, which consists of silica, is stable in temperate climates at the surface.

6. Describe the different horizons in a typical soil profile.

    Nearest to the surface, the O-horizon consists largely of humus, partially decayed organic matter. Below the O lies the A-horizon, in which humus is mixed with clay, silt, and sand grains. The upper part of A (termed A_1) is darker than the lower part (termed A_2) and contains proportionately more humus and less clastic material; together, the O- and A_1-horizons form the vernacular "topsoil." Beneath A, the B-horizon consists of clay and minerals (including iron oxides) derived from the overlying layers and transported (either as detritus or dissolved ions) by percolating groundwater. The C-horizon below B consists of chemically altered bedrock or sediment, which grades into unaltered bedrock or sediment.

7. What factors affect the nature of soils in particular regions?

Temperature, precipitation rate, and the nature of the substrate (underlying bedrock or sediment) are the predominant factors that affect soil character.

8. Describe how a sedimentary rock is formed from its unweathered parent rock.

First, physical and chemical weathering break up and alter the parent rock to form detrital fragments of parent material, dissolved ions, and clay. This material is then eroded from the parent surface and transported away from the source by water, wind, or glacial ice (or directly by gravity, in the case of large clasts on a slope). Ultimately, the detritus will settle out of the transport medium and the dissolved ions may react to form chemical precipitates (deposition includes both physical settling and chemical precipitation). Eventually, loose grains of deposited sediment may become buried under additional sediment, compacted, and cemented to form sedimentary rock.

9. Clastic and chemical sedimentary rocks are both made of material that has been transported. How are they different?

The materials that form chemical rocks are transported as dissolved ions, which precipitate to form solid minerals at the site of deposition. Clastic debris is carried as solid detritus, which deposits by physically settling out of its transport medium.

10. Is a rock made of shell fragments a clastic or a chemical sedimentary rock? Explain your reasoning.

All limestones are chemical sedimentary rocks in the sense that their mineral components (calcite or aragonite) arise from chemical precipitation. Most often, these precipitation reactions are biologically catalyzed, and these limestones are often placed in a biochemical subcategory. As noted in the text, clastic (detrital) rocks are those derived from detritus weathered and eroded from preexisting rock. Thus, a limestone composed of shell fragments is not a clastic rock, it is a (bio)chemical rock with a clastic texture (consisting of cemented fragments). Occasionally, such limestones are termed bioclastic, but it is recommended here that "clastic" be reserved for lithified detritus derived from physical weathering of rock.

However, if a body of limestone were to be subaerially exposed at some point in the future, it could conceivably, in a cold, dry climate, physically weather to form detritus which might lithify to form a clastic rock. (Limestone is highly susceptible to dissolution through chemical weathering that dominates in warm, moist settings.)

11. Describe how grain size and shape, sorting, sphericity, and angularity change as sediments move downstream.

Mechanical forces such as tumbling and abrasion wear on sediments as they are transported downstream. Angular protuberances are especially likely to be broken off. As a result, grain size decreases, with grains becoming more spherical and more rounded (less angular). The speed at which the water in a stream is traveling decreases along its course, and the capacity of the stream to carry sediment is directly related to its rate of flow. So larger grains are deposited upstream from finer grains (grains become sorted as they travel downstream).

12. Describe the two different kinds of chert. How are they similar? How are they different?

All chert is composed of microcrystalline (cryptocrystalline) quartz. Biochemical chert is derived from the siliceous skeletons of microorganisms (diatoms and radiolaria), which deposit in vast layers on the sea floor after the death of the organisms. Chert also occurs as a replacement mineral; groundwater may dissolve portions of limestone (when groundwater is undersaturated with respect to calcite) and fill in the resultant void space with microcrystalline quartz (with respect to which the groundwater is saturated). Bedding (layering) is absent in replacement chert.

13. What kinds of conditions are required for the formation of evaporites?

Evaporite formation requires a fluid with dissolved ions (typically seawater) to be evaporated to such an extent that the ions will precipitate out, typically as halide, carbonate, or sulfate minerals. Hot, dry conditions with extensive subaerial exposure are conducive to evaporite formation; these include warm, broad, shallow seas with little riverine influx and restricted circulation. Along coasts in regions with a hot, dry climate, evaporites are deposited in the supratidal (above the high tide limit) "splash zone" where water washes up, becomes stranded (having lost contact with the cohesive ocean), and is evaporated by the heat of the Sun. Similarly, on the dry portions of continents, such as the western United States, isolated saline lakes and ephemeral streams produce evaporates when their water levels fall.

14. What minerals precipitate out of seawater first? next? last? What does this suggest when geologists find huge volumes of pure gypsum in the Earth's crust?

Carbonate minerals (calcite, dolomite) precipitate first, sulfates (gypsum, anhydrite) are next, and halides (halite, sylvite) are last. The carbonates, however, are produced in much smaller quantities, so that most evaporate sequences are predominantly gypsum and halite. The presence of pure gypsum would suggest that halite from the evaporite sequence has been redissolved subsequent to deposition.

15. How is dolostone different from limestone?

Dolostone is composed of the mineral dolomite $(Ca, Mg)CO_3$, whereas limestone is composed of calcite $CaCO_3$. Dolostone contains more magnesium (a trace impurity in natural calcite) and has a different crystalline structure.

16. Describe how cross beds form. How can you read the current direction from cross beds?

Cross beds form from sediment deposited on the lee side of dunes and ripples. These angled beds dip downward in the downcurrent direction (parallel to the lee side of the dune or ripple).

17. Describe how a turbidity current forms and moves. How does it produce graded bedding?

Turbidity currents form when sediment becomes unstable on a subaqueous slope and tumbles downward, pulling a current of water with it. After a while, the velocity of the turbidity current slows. The heaviest (largest) particles settle out first, whereas smaller, lighter particles stay in suspension for longer periods. Ultimately these smaller particles settle atop the coarser grains to produce graded bedding, a grain-size gradient from coarse (near the base of the bed) to fine (at the top of the bed).

18. Compare the geometry and typical sediments of an alluvial fan with a typical river environment and with a deep-marine deposit.

Alluvial fans are wedge-shaped deposits occurring at the foot of an eroding mountain range. The sediments are typically coarse (sand, pebbles, or cobbles) and contain substantial amounts of feldspar (physically weathered from a typically granitic montane source).

Rivers produce clastic deposits dominated by quartz and clay minerals. Laterally, sheets of mud form floodplain deposits. The river channel itself is represented by a lens-shaped body of rippled, cross-bedded sand.

Deep-marine deposits are dominated by the skeletons of planktonic microorganisms (chalk derived from foraminiferans and bedded cherts derived from diatoms and radiolaria) and clay (which settles to form finely laminated mudstone).

19. Why don't sediments accumulate everywhere? What kind of tectonic conditions are required to create basins?

In a majority of terrestrial environments, the rate of erosion meets or exceeds the rate of sedimentary deposition. Basin subsidence allows for the development and preservation of thick sedimentary sequences (including nonmarine deposits) but requires a locally sinking lithosphere. Sinking lithosphere is most often a consequence of tectonic rifting or collision, when the lithosphere is either stretched and thinned or subjected to a load.

Test bank

1. The breakdown of exposed rock into small fragments and dissolved ions is termed _____.
   A. deposition	B. erosion	C. weathering

2. The removal of detritus from weathered rock at an outcrop is termed _____.
   A. deposition	B. erosion	C. weathering

3. The majority of the rocks that occur at the surface of the Earth are _____.
   A. intrusive igneous rocks	B. extrusive igneous rocks
   C. sedimentary rocks	D. metamorphic rocks

4. Geologically, medium-sized sediment refers to _____.
   A. cobbles	B. pebbles	C. sand	D. silt

5. Frost wedging, root wedging, and salt wedging are all examples of _____.
   A. erosion	B. chemical weathering
   C. physical weathering	D. deposition

6. Hydrolysis, oxidation, and hydration are all examples of _____.
   A. erosion	B. chemical weathering
   C. physical weathering	D. deposition

7. Chemical weathering takes place most rapidly in environments that are _____.
   A. cool and dry	B. cool and wet
   C. warm and dry	D. warm and wet

8. Rapid physical weathering _____.
   A. does not affect the rate of chemical weathering in the same rock
   B. tends to reduce the rate of chemical weathering in the same rock
   C. tends to increase the rate of chemical weathering in the same rock

9. Rapid chemical weathering _____.
   A. does not affect the rate of physical weathering in the same rock
   B. tends to reduce the rate of physical weathering in the same rock
   C. tends to increase the rate of physical weathering in the same rock

10. A pile of rubble at the base of an outcrop derived from weathered rock upslope is geologically termed _____.
    A. rubbish	B. talus
    C. spoilage	D. decollement

11. The chemical reaction that transforms feldspar into clay is an example of _____.
    A. hydrolysis	B. hydration
    C. dissolution	D. oxidation

12. The rusting of iron and iron-rich minerals is an example of _____.
    A. hydrolysis   B. hydration
    C. dissolution   D. oxidation

13. The swelling of certain minerals due to the incorporation of water into their crystal lattices is termed _____.
    A. hydrolysis   B. hydration
    C. dissolution   D. oxidation

14. On Earth, loose sediment that covers bedrock and has been chemically altered by reactions with rainwater and the addition of organic matter by the biosphere is termed _____.
    A. soil   B. loess   C. regolith   D. caprock

15. On the Moon, loose sediment covering bedrock is termed _____.
    A. soil   B. loess   C. regolith   D. caprock

16. Topsoil consists of which soil horizon(s)?
    A. A-horizon   B. B-horizon
    C. C-horizon   D. O-horizon and the upper portion of the A-horizon

17. Which soil horizon is the zone of accumulation, so named because dissolved matter, leached from other parts of the soil, precipitates to form new minerals.
    A. A-horizon   B. B-horizon
    C. C-horizon   D. O-horizon

18. Which soil horizon is the uppermost?
    A. A-horizon   B. B-horizon
    C. C-horizon   D. O-horizon

19. Which soil horizon is chemically most similar to the underlying bedrock or unaltered sediment?
    A. A-horizon   B. B-horizon
    C. C-horizon   D. O-horizon

20. Which soil horizon has the greatest proportion of organic matter?
    A. A-horizon   B. B-horizon
    C. C-horizon   D. O-horizon

21. Caliche is most commonly found in which type of environment?
    A. temperate forests   B. tropical rain forests
    C. deserts   D. grasslands

22. Laterite soils are most commonly found in which type of environment?
    A. temperate forests   B. tropical rain forests
    C. deserts   D. grasslands

23. Lithified detritus (breakdown products of preexisting rocks) forms which kind of sedimentary rock?
   A. biochemical			B. chemical
   C. clastic			D. organic

24. Cemented shells of marine organisms form which kind of sedimentary rock?
   A. biochemical		B. clastic		C. organic

25. Physical precipitation of gypsum due to evaporation of seawater produces which kind of sedimentary rock?
   A. biochemical			B. chemical
   C. clastic			D. organic

26. Clastic sedimentary rocks are primarily classified on the basis of _____.
   A. grain size			B. degree of sorting
   C. angularity			D. mineral composition

27. If water is the transport medium of sediment, the grain size of sedimentary deposits most closely indicates the _____.
   A. geographic extent of the weathering source rock at outcrop
   B. average velocity of the water from the time of erosion until deposition
   C. velocity of the water at the moment the sediment settled to the bottom

28. Compaction and cementation of grains occurs during _____.
   A. erosion			B. lithification
   C. transport			D. weathering

29. Which transport medium carries the largest particles?
   A. ice		B. water		C. wind

30. It is unusual for _____ to carry grains larger than sand.
   A. ice		B. water		C. wind

31. The difference between breccia and conglomerate is that conglomerate _____.
   A. is finer-grained than breccia
   B. is coarser-grained than breccia
   C. possesses more angular grains than breccia
   D. possesses more rounded grains than breccia

32. Compared to arkose, quartz sandstone _____.
   A. is more mature
   B. does not contain significant amounts of feldspar
   C. is likely to be found farther away from weathering granitic source rock
   D. All of the above are correct.

33. Mud, sand, and lithic clasts make up sedimentary rock termed _____.
   A. metasandstone			B. lithosandstone
   C. graywacke			D. arkose

34. A clastic rock made up of sand-sized grains derived from the physical weathering of granite and containing a sizable proportion of feldspar is termed _____.
   A. metasandstone B. lithosandstone
   C. graywacke D. arkose

35. A fine-grained clastic rock that splits into thin sheets is _____.
   A. mudstone B. shale
   C. sandstone D. arkose

36. Because the velocity of sediment settling (deposition) is positively related to grain size for waterborne sediments, fluvial deposits are more likely than glacial deposits to _____.
   A. be well sorted
   B. include coarse grains, such as cobbles
   C. include fine grains, such as clay
   D. have angular grains

37. Of the grains above, which possesses the greatest sphericity?
   A. grain A
   B. grain B
   C. Both grains possess approximately equal sphericity.

38. Of the grains above, which is most rounded?
   A. grain A
   B. grain B
   C. Both grains possess approximately equal roundness.

39. An arkose with well-rounded, well-sorted sand grains of feldspar and quartz is said to be _____.
   A. physically and chemically mature
   B. physically mature but chemically immature
   C. chemically mature but physically immature
   D. physically and chemically immature

40. A breccia consisting almost entirely of quartz and clay minerals is said to be _____.
   A. physically and chemically mature
   B. physically mature but chemically immature

60

C. chemically mature but physically immature
D. physically and chemically immature

41. Well sorted fine sandstone composed nearly entirely of quartz is said to be _____.
    A. physically and chemically mature
    B. physically mature but chemically immature
    C. chemically mature but physically immature
    D. physically and chemically immature

42. Chemical and biochemical rocks are classified primarily on the basis of _____.
    A. grain size   B. degree of sorting
    C. angularity   D. mineral composition

43. Biochemical limestones are dominated by carbonate mud and fragments of _____.
    A. siliceous shells of planktonic diatoms and foraminifera
    B. calcitic and aragonitic skeletons of marine invertebrates
    C. the phosphatic bones of fish
    D. the organic breakdown products of wood from trees

44. Unlike physically precipitated chert, biochemical chert _____.
    A. is bedded (layered)
    B. is made up of the skeletons of radiolaria and foraminifera
    C. does not occur as small nodules within bodies of limestone
    D. All of the above are correct.

45. Flint, once commonly used to form arrowheads, is a _____.
    A. red variety of biochemical limestone
    B. black variety of biochemical chert
    C. black variety of replacement chert
    D. black variety of chemical limestone

46. Physically precipitated (chemical) limestone that forms in caves is termed _____.
    A. agate   B. dolostone   C. jasper   D. travertine

47. When limestone becomes chemically altered so that half of the calcium atoms are replaced by magnesium, the resultant rock is termed _____.
    A. agate   B. dolostone   C. jasper   D. travertine

48. Stratification refers to _____.
    A. the development of layering within sedimentary rocks
    B. the act of deposition of sediment, which will ultimately form sedimentary rock
    C. physical and chemical alterations, including compaction and cementation, that occur as sediment is transformed into rock

49. Diagenesis refers to _____.
   A. the development of layering within sedimentary rocks
   B. the act of deposition of sediment, which will ultimately form sedimentary rock
   C. physical and chemical alterations, including compaction and cementation, that occur as sediment is transformed into rock

50. Sea level rises locally and marine sediments are deposited on top of terrestrial sediments during events termed _____.
   A. regressions
   B. transgressions
   C. divarications

Chapter 8
Change in the Solid State: Metamorphic Rocks

Learning objectives

1. Students should know the conditions under which metamorphic rocks form and that chemical weathering and low-temperature diagenesis, although processes of chemical alteration, are not included within metamorphism.

2. They should be aware of the three distinct settings for metamorphism: thermal (contact), dynamic (fault zone), and dynamothermal (regional).

3. Metamorphism is induced by heat, pressure, and differential stress. Heat can be added by burial depth or contact with hot groundwater. Pressure is added with depth. Differential stress arises at fault zones and over broad regions during orogenesis.

4. Metamorphic effects include recrystallization, metasomatism, changes in mineral stability, and foliation (metamorphic development of planar fabric). Foliation involves the development of preferred orientation (of inequant grains) or preferred mineral associations (compositional banding).

5. Students should be familiar with a couple of nonfoliated metamorphic rocks (quartzite, marble) and the slate-phyllite-schist-gneiss series of foliated rocks and know the type of foliation present in each member of the series.

6. Temperature-pressure fields are termed metamorphic facies. Facies are diagnosed from rocks using index minerals and mineral assemblages. High-grade rocks are formed at high temperatures (and pressures); low-grade rocks are produced at lower temperatures (and pressures).

7. Protoliths affected by metamorphism can be of any type (igneous, sedimentary, or metamorphic). Alteration of metamorphics to produce higher-grade rocks is termed prograde metamorphism; the reverse process is termed retrograde metamorphism.

Summary from the text

Metamorphism refers to changes in a rock that result in the formation of a metamorphic mineral assemblage and/or a metamorphic foliation without the rock melting or becoming sediment. The new rock that results from these changes is a metamorphic rock.

Metamorphic mineral assemblages form when the original minerals in a protolith become unstable owing to pressure and temperature changes, and recrystallization rearranges atoms into new mineral crystals. If hot-water solutions bring in or remove atoms, we say that metasomatism has occurred.

Geologists separate metamorphic rocks into two classes, foliated rocks and nonfoliated rocks, depending on whether the rock contains foliation.

Metamorphic foliation can be defined either by compositional banding or by a preferred mineral orientation (aligned inequant crystals). Compositional banding forms when light minerals separate from dark minerals during recrystallization. Preferred mineral orientation develops where differential stress causes the squashing and shearing of a rock, so that its inequant grains align parallel with each other.

The class of foliated rocks includes slate, metasandstone and metaconglomerate, phyllite, schist, amphibolite, gneiss, and migmatite; the class of nonfoliated rocks includes hornfels and some kinds of quartzite and marble.

Metamorphic conditions are the temperatures and pressures necessary to cause metamorphism. Temperature is the more important factor. Rocks formed under relatively low temperatures are known as low-grade rocks, while those formed under elevated temperatures are known as high-grade rocks. Intermediate-grade rocks develop under intermediate conditions. Slate and phyllite are low grade, most schist is intermediate grade, and most gneiss and all migmatite are high grade.

Geologists track the distribution of different grades of rock by looking for index minerals, which indicate the temperature and pressure at which a rock formed. We then can map out metamorphic zones, regions in which an index mineral occurs, and isograds, the boundary lines between zones.

A metamorphic facies is a group of metamorphic rocks that develops under a specified range of temperature and pressure conditions (more precisely defined than those indicated by grade). Facies are recognized by the occurrence of an assemblage of minerals; the assemblage in a given rock depends on the composition of the protolith as well as the metamorphic conditions.

Thermal metamorphism (also called contact metamorphism) occurs in an aureole surrounding an igneous intrusion. Because there is no shearing involved, nonfoliated rocks form in contact aureoles. Dynamically metamorphosed rocks form along faults, where rocks are only sheared, under metamorphic conditions. Dynamothermal metamorphism (also called regional metamorphism) results when rocks are buried deeply during mountain building. Because such metamorphism involves shearing and squashing as well as heat, rocks develop foliation.

Metamorphism occurs because of plate interactions: the process of mountain building in convergent zones causes dynamothermal metamorphism; shearing along plate boundaries causes dynamic metamorphism; and igneous plutons in rifts cause thermal metamorphism. The circulation of hot water causes a retrograde metamorphism of oceanic crust at mid-ocean ridges. Unusual metamorphic rocks called blueschists form at the base of accretionary prisms, where pressures are high but temperatures are low.

We find extensive areas of metamorphic rocks in mountain ranges. Vast regions of continents known as shields expose ancient (Precambrian) metamorphic rocks.

Answers to review questions

1. How are metamorphic rocks different from igneous and sedimentary rocks?

Metamorphic rocks are the result of heat and pressure causing an alteration of texture, mineralogy, or both within a preexisting rock, without the rock having undergone melting or surficial weathering. Many metamorphic (but no igneous or sedimentary) rocks possess foliation (discussed below).

2. What two features characterize most metamorphic rocks?

Metamorphic mineral assemblages (minerals uniquely produced under the temperature and pressure regimes of metamorphism) and metamorphic foliation (layering or preferred grain orientation resultant from metamorphic pressure) are characteristic of most metamorphic rocks.

3. How do heat, hot groundwater, and pressure change a rock?

Changes in temperature and pressure favor chemical reactions that produce minerals that would not otherwise form. Hot groundwater can be an agent of metasomatism (net chemical change in a body of rock) through cation exchange and may also deposit minerals into open cracks, forming veins.

4. How does differential stress determine mineral orientation?

Under sufficiently differential stress, inequant (platy or elongate) grains can be rotated into preferred orientation, become deformed until they attain a preferred orientation, or grow in a preferred orientation.

5. What is foliation?

Foliation is the metamorphic production of a planar fabric in a rock, which may consist of the growth of compositional bands, the deformation of grain shapes, or the development of a preferred orientation of inequant grains.

6. How is a slate different from a phyllite? How does a phyllite differ from a schist?

Slate and its characteristic slaty cleavage arise from the preferred orientation of clay minerals resulting from the relatively low-temperature and low-pressure metamorphism of a body of slate. Phyllite arises when significantly higher temperatures and pressures cause clay grains within slate to be recrystallized to form mica grains, which retain a preferred orientation. Unlike slate, which is rather dull, mica gives phyllite a silky luster. Schist differs from phyllite in that, as a result of greater heat and pressure, the mica grains are large, visible discrete plates, unlike the smooth sheen of tiny mica grains within phyllite.

7. Why are quartzites and marbles usually nonfoliated?

Quartzite and marble result from metamorphism of sandstone and limestone, respectively. Both sandstone and limestone are effectively monomineralic (being composed of quartz and calcite, respectively) so compositional banding cannot develop, and neither contains any mica or any substantial quantities of minerals with compositions similar to mica. Their component quartz and calcite occur in equant grains, so preferred orientations cannot be observed.

8. What is a metamorphic grade? How is it recognized in a rock?

A metamorphic grade refers to a series of temperature and pressure regimes under which metamorphism takes place. For example, high-grade metamorphism occurs under greater temperatures (and pressures) than low-grade metamorphism. Metamorphic grade is usually assessed on the basis of the mineral assemblage making up the metamorphic rock, as well as its foliation and other textural clues (such as grain size).

9. How does prograde metamorphism differ from retrograde metamorphism?
   Prograde metamorphism occurs when a low-grade (produced at relatively low temperature and pressure) metamorphic rock is subjected to increased temperature and pressure to form a higher-grade metamorphic rock. Retrograde metamorphism is the reverse process, wherein decreasing temperature and pressure change a metamorphic rock into a lower-grade metamorphic rock.

10. Describe the geologic settings where thermal, dynamic, and dynamothermal metamorphism take place.
    Thermal metamorphism takes place in a zone of country rock surrounding a pluton, where the country rock's mineral assemblage becomes recrystallized. Dynamic metamorphism occurs in fault zones, where shearing force recrystallizes minerals at depth. Dynamothermal metamorphism occurs within the cores of mountain ranges, induced by increase heat and pressure associated with crustal thickening and the shear that arises in the development of fold-and-thrust belts.

11. How does plate tectonics explain the peculiar combination of low-temperature but high-pressure minerals found in a blueschist?
    Blueschists form at the base of thick accretionary prisms, sediments scraped off of the downgoing slab at subduction zones. Because the subducting slab is relatively cool, it adds little heat to the prism, allowing for the relatively high pressures but low temperatures in which the blueschist mineral assemblage is stable.

Test bank

1. A body of gneiss is subjected to heat and forms a melt. Later the melt cools and crystallizes to form a(n) _____.
   A. metamorphic rock
   B. igneous rock
   C. sedimentary rock

2. Metamorphism brings changes in mineral arrangement and the texture of rocks, but it never leads to new mineral assemblages.
   A. true   B. false

3. A buried body of aragonitic limestone is recrystallized at low temperatures and pressures, producing calcite; this is an example of _____.
   A. diagenesis   B. erosion
   C. metamorphism   D. weathering

4. A buried body of shale is subjected to differential stress, causing clay minerals to realign and producing slate. This is an example of _____.
   A. diagenesis   B. erosion
   C. metamorphism   D. weathering

5. At the surface, potassium feldspar reacts with water to form clay; this is an example of _____.
   A. diagenesis		B. erosion
   C. metamorphism	D. weathering

6. Clay minerals within a buried body of slate are recrystallized at high temperatures and pressures to form mica, producing a rock called phyllite; this is an example of _____.
   A. diagenesis		B. erosion
   C. metamorphism	D. weathering

7. Metamorphism may be induced by _____.
   A. contact with a hot pluton
   B. contact with hot groundwater
   C. heat and pressure associated with deep burial
   D. All of the above are correct.

8. Differential stress will cause crystals to align in a preferred orientation unless the crystals are _____.
   A. primarily mica	B. platy
   C. equant		D. elongate

9. Squashing a fly with a fly swatter is an application of _____.
   A. normal stress	B. shear stress

10. Spreading peanut butter on bread is an application of _____.
    A. normal stress	B. shear stress

11. A primary difference between quartzite and metasandstone is _____.
    A. metasandstone is dark brown whereas quartzite is white to tan
    B. quartzite is produced through dynamothermal metamorphism but metasandstone is not
    C. quartzite is foliated, but metasandstone generally is not
    D. metasandstone is foliated, but quartzite generally is not

12. A primary difference between phyllite and schist is _____.
    A. schist contains mica, but phyllite contains only clay
    B. phyllite contains mica, but schist contains only clay
    C. mica crystals within schist are larger than those within phyllite
    D. mica crystals within phyllite are larger than those within schist

13. The development of a preferred orientation of large, flaky mica crystals within metamorphic rock is termed _____.
    A. slaty cleavage	B. phyllitic luster
    C. schistosity	D. compositional banding

14. Slaty cleavage, schistosity, and compositional banding are all examples of _____.
    A. mineral cleavage   B. foliation
    C. recrystallization   D. sedimentary structures

15. A compositionally banded rock with a mineral assemblage similar to granite is _____.
    A. granodiorite   B. granitic gneiss
    C. phyllite   D. slate

16. Two common metamorphic rocks that typically lack foliation are _____.
    A. slate and phyllite   B. gneiss and migmatite
    C. quartzite and marble   D. schist and metaconglomerate

17. Compared to low-grade metamorphic rocks, high-grade rocks _____.
    A. always contain more quartz and feldspar
    B. are produced closer to the surface, high in the stratigraphic column
    C. are produced at greater temperatures and pressures
    D. are produced at cooler temperatures, but greater pressures

18. A mineral within a metamorphic rock that can be used to provide a narrow constraint on the temperature and pressure of formation of the rock is termed a(n) _____.
    A. thermineral   B. index mineral
    C. mafic mineral   D. halide mineral

19. The process of low-grade metamorphic rocks being altered to form high-grade metamorphic rocks is termed _____ metamorphism.
    A. foliated   B. prograde   C. retrograde   D. dynamic

20. The process of high-grade metamorphic rocks being altered to form low-grade metamorphic rocks is termed _____ metamorphism.
    A. foliated   B. prograde   C. retrograde   D. dynamic

21. Alteration due uniquely to the shear encountered at a fault zone is termed _____ metamorphism.
    A. foliated   B. prograde   C. retrograde   D. dynamic

22. The region of thermally metamorphosed rock surrounding a cooled pluton is called a(n) _____.
    A. shear zone   B. aureole   C. oriole   D. oleo

23. Rocks resulting from thermal metamorphism will not possess _____.
    A. a new mineral assemblage distinct from that found prior to intrusion
    B. larger crystals than those characterizing the country rock prior to intrusion
    C. foliation
    D. silicate minerals

24. Regional metamorphism _____.
    A. takes place at cool temperatures and low pressure
    B. takes place at cool temperatures but high pressure
    C. is another name for thermal metamorphism
    D. is another name for dynamothermal metamorphism

25. Which type of metamorphism affects the greatest volumes of rock?
    A. thermal metamorphism
    B. dynamothermal metamorphism
    C. dynamic metamorphism

26. The protolith subjected to metamorphism _____.
    A. is always metamorphic rock to begin with
    B. is always igneous rock
    C. is always sedimentary rock
    D. may belong to any of the three primary rock types

27. The blueschist facies is a metamorphic realm of _____.
    A. high temperature and pressure
    B. low temperature and pressure
    C. high temperature but relatively low pressure
    D. high pressure but relatively low temperature

28. Within a single mountain range, _____.
    A. only low-grade metamorphic rocks are likely to be found
    B. only high-grade metamorphic rocks are likely to be found
    C. it is possible to find a variety of metamorphic rocks produced in distinct facies, including high-, low-, and intermediate-grade rocks

29. Precambrian metamorphic rocks are exposed at the surface _____.
    A. on Mars and Venus, but nowhere on Earth
    B. at places in continental interiors termed platforms
    C. at places in continental interiors termed shields
    D. at the bottom of the deep sea

30. Net chemical change in metamorphic rock induced by reaction with hot groundwater is termed _____.
    A. foliation   B. metasomatism   C. anachronism

31. The mineral assemblage within metamorphic rock is _____.
    A. always identical to that found within the protolith
    B. dependent only upon the mineral assemblage of the protolith
    C. dependent only upon the temperature and pressure of formation
    D. dependent upon both the mineral content of the protolith and the temperature and pressure of formation

32. The blue mineral for which the blueschist facies is named is _____.
    A. smoky blue quartz   B. topaz
    C. glaucophane   D. talc

33. Compared to the amphibolite metamorphic facies, the greenschist facies _____.
    A. consists of lower-grade rocks
    B. consists of higher-grade rocks
    C. is an identical temperature and pressure regime; greenschists and amphibolites bear different mineral assemblages only because of differences in protolith chemistry

34. Thermal (contact) metamorphism _____.
    A. occurs in areas surrounding igneous intrusions
    B. occurs only where gneiss is in contact with schist
    C. occurs as a consequence of the sinking of a broad region to great depth
    D. only takes place at the surface, where rock is in contact with the atmosphere

35. Dynamothermal metamorphism occurs when _____.
    A. rock becomes buried deeply during continental collision and mountain building
    B. regression of the sea leads to erosion of sedimentary cover atop a body of rock
    C. the upper surface of a body of rock develops a thick soil profile
    D. a pluton causes metamorphism in a small surrounding region

36. Which of the following processes cannot occur in the formation of metamorphic rock?
    A. realignment of minerals so that they develop a preferred orientation
    B. segregation of minerals into layers of different compositions
    C. solid-state rearrangement of atoms or ions to create a new assemblage of minerals
    D. complete re-melting of the rock, followed by solidification to form a new rock

37. Metamorphism, in broadest terms, involves _____.
    A. the settling of crystals in a melt as it cools
    B. the sorting of grains by size, as is accomplished by rivers and beach waves
    C. cementation of loose grains and precipitation of new minerals into pore spaces
    D. changes in mineralogy and texture in response to heat and pressure

38. Gneiss typically forms under higher pressures than does hornfels.
    A. true   B. false

39. Which list properly orders metamorphic rocks from lowest to highest grade?
    A. conglomerate, sandstone, siltstone, shale
    B. shale, slate, phyllite, quartzite
    C. slate, phyllite, schist, gneiss
    D. gneiss, phyllite, schist, slate

40. Foliated metamorphic rocks possess _____.
    A. leafy plant fossils (ancient foliage)
    B. a homogenous texture resulting from randomly oriented grains
    C. a planar fabric consisting of mineral grains in preferred orientations or preferred patterns of association (banding)
    D. minerals precipitated directly from seawater

Chapter 9
The Wrath of Vulcan: Volcanic Eruptions

Learning objectives

1. Students should be familiar with the three primary volcanic products: lava, gas, and pyroclastic debris. They should know why some eruptions are dominated by lava flow, others by pyroclastic debris, and that each type of eruption can occur within a single volcano at different times.
2. They should know the three types of volcanoes: cinder cones, shield volcanoes, and stratovolcanoes and how each is produced.
3. Students should be familiar with the mineralogically based classifications of lavas (similar to igneous rocks), and they should know the factors that affect lava viscosity. Similarly, they should know that basaltic lavas erupt at higher temperatures than rhyolitic lavas and are not associated with explosive eruption as rhyolitic and intermediate melts are.
4. They should know the variety of pyroclastic debris products (ash, cinders, blocks, bombs, tephra, tuff, ignimbrite), and the variety of dangers (earthquakes, landslides, lahars) associated with volcanic eruptions.
5. Volcanoes are common at convergent and divergent plate boundaries and generally absent at transform boundaries, but they may arise in plate interiors in association with a hot mantle plume (hot spot).
6. Earth is not unique in possessing volcanoes. The largest known is an extinct shield volcano, Olympus Mons, found on Mars. The most volcanically active body in the solar system is Io, a moon of Jupiter.

Summary from the text

Volcanoes are vents at which molten rock (lava), pyroclastic debris (ash, pumice, and fragments of volcanic rock), and gas erupt at the Earth's surface. A hill or mountain created from the products of an eruption is also called a volcano.

The characteristics of a lava flow depend on its viscosity, which in turn depends on its temperature and composition. Silicic (rhyolitic) lavas tend to be more viscous than mafic (basaltic) lavas.

Basaltic lavas can flow great distances. Pahoehoe flows have smooth, ropy surfaces, while aa flows have rough, rubbly surfaces. In some cases, columnar joints form in a lava flow when it cools.

Andesitic and rhyolitic lava flows tend to pile into mounds at the vent.

Pyroclastic debris includes powder-sized ash, marble-sized lapilli, and basketball- to refrigerator-sized blocks. Frothy lava freezes to form sponge-like pumice, and cemented ash makes up tuff.

Fast-moving pyroclastic flows (glowing avalanches) solidify into a rock called ignimbrite.

Volcanic eruptions may emit many kinds of gas. Gas bubbles frozen into rock are vesicles.

Eruptions may occur at a volcano's summit or from fissures on its flanks. The summit of an erupting volcano may collapse to form a bowl-shaped depression called a caldera.

A volcano's shape depends on the type of eruption. Shield volcanoes are broad, gentle domes. Cinder cones are steep-sided, symmetrical hills composed of tephra. Composite volcanoes can become quite large and consist of alternating layers of pyroclastic debris and lava.

The type of eruption depends on the lava's viscosity and gas content. Effusive eruptions produce only flows of lava, while explosive eruptions produce clouds and flows of pyroclastic debris.

Volcanoes form in a variety of plate tectonic settings: oceanic hot spots, continental hot spots, mid-ocean ridges, subduction zones, and rifts. Different kinds of volcanoes develop in these settings.

Volcanic eruptions pose many hazards: lava flows overrunning roads and towns, ash falls blanketing the landscape, pyroclastic flows incinerating towns and fields, landslides and lahars burying valleys, earthquakes toppling structures and rupturing dams, tsunamis washing away coastal towns, and invisible gases suffocating nearby people and animals.

Geologists distinguish between active, dormant, and extinct volcanoes on the basis of the likelihood that the volcano will erupt.

Eruptions can be predicted through changes in heat flow, changes in shape, earthquake activity, and the emission of gas and steam.

We can minimize the consequences of an eruption by avoiding construction in danger zones (such as the path of lahars) and by drawing up evacuation plans. In a few cases, it may be possible to divert flows.

Volcanic gases and ash, erupted into the stratosphere, may keep the Earth from receiving solar radiation and thus may affect climate. Eruptions also bring nutrients from inside the Earth to the surface, and eruptive products may evolve into fertile soils.

The largest known volcano in the solar system, Olympus Mons, towers over the surface of Mars. Satellites have documented eruptions on Io, a moon of Jupiter.

Answers to review questions

1. Describe the three different kinds of material that can erupt from a volcano.

Lava is molten rock, flowing out of the volcano in liquid state and cooling and solidifying on the ground. Pyroclastic debris forms when bodies of lava are shot into the atmosphere and cool during flight to form volcanic glass; pyroclastic debris ranges in size from ash to cinders (lapilli) to larger pieces termed blocks. Being larger, blocks take longer to cool than does ash; as they hurtle through the atmosphere they are commonly molded into aerodynamic shapes called volcanic bombs. Lastly, the primary gases that erupt from volcanoes are water vapor, carbon dioxide, sulfur dioxide, and hydrogen sulfide.

2. How does a lava tube form?

A lava flow does not solidify all at once; for example, the uppermost surface is in contact with the cooling air and freezes before the interior of the flow will. Hotter segments of the lava flow fastest, reducing local pressure; this attracts more molten lava, so that it is common for "pipelines" of molten lava to flow deep within bodies of lava that are otherwise solidified. When the eruption of lava is complete, the lava that remains in a molten state the longest may flow out of the pipeline and beyond the edge of the main flow before solidifying, leaving behind an empty lava tube made up of the solidified material that immediately surrounded the pipeline lava.

3. How does a pillow lava form?

Pillow lavas form during submarine eruptions, such as those at mid-ocean ridges. Water cools and solidifies lava far more quickly than does air, so erupted bodies of subaqueous lava do not flow very far before solidifying, and the extensive, sheet-like flows associated with continental volcanism cannot develop. However, pillows occur not singly but rather in genetically associated chains. The outer surface of the initial pillow, in contact with the water, solidifies while the interior is still molten and inclined to flow. Eventually, pressure from the molten filling, still receiving lava from the volcanic source, causes the leading edge of the solidified rind to burst. Lava then flows for a short distance before another "pillowcase" rind forms and temporarily impedes the flow of lava, only to be broken, recursively forming another short flow and, farther down, yet another pillow. The cycle of pillow and flow is broken when the pillows freeze solid and new lava flows over the surface of the extant pillows to create a new pillow layer.

4. Describe the differences between a pyroclastic flow and a lahar.

Pyroclastic flows are hot mixtures of suspended pyroclastic debris (ash) and air that tumble down the sides of volcanoes at great speed. A lahar is a fast, liquid flow arising when a pyroclastic flow mixes with water from snow fields or nearby streams or rivers.

5. How is a crater different from a caldera?

A crater is a relatively smaller basin (less than 500 m in diameter) at the summit of a volcanic cone, formed from lava and ash around the margins of the chimney. From the aperture, the walled crater may extend to a depth of 200 m. An explosive or otherwise large eruption may cause the walls of the cone to collapse upon the (now empty) magma chamber to form a larger circular depression termed a caldera. Calderas are flat floored, with diameters of up to thousands of meters. They may reach depths of hundreds of meters.

6. Describe the differences between shield volcanoes, stratovolcanoes, and cinder cones. How are these differences explained by the composition of their lavas and other factors?

Shield volcanoes are gently sloped domes typically composed of basaltic volcanic rock. The lava that forms this rock is of low viscosity, so it flows readily in response to gravity and attains a low profile upon solidification. Cinder cones are radially symmetric tephra, with steeper slopes on the sides. Cinder cones form from fountains of lava, which squirt up and freeze in midair close to the volcanic vent. Pyroclastic tephra solidifies before reaching the ground and consequently cannot flow as basaltic lava will. Stratovolcanoes, or composite volcanoes, are large, conic volcanoes. The walls of stratovolcanoes are composed of alternating layers of tephra and lava rock. Stratovolcanoes are the result of multiple eruptions, which have in turns produced both effusive lava and explosive pyroclastics, whereas the smaller cinder cones represent single volcanic events.

7. Why do some volcanic eruptions consist mostly of lava flows, while others are explosive and have no flow?

Explosive eruptions are the result of a sudden release of accumulated gas pressure within the volcano. Lavas with high proportions of volatiles, such as water vapor, produce explosive eruptions in which the lava is blasted upward by explosively expanding gas, forming pyroclastic debris out of the lava that would otherwise have flowed down the slope of the volcano.

8. Explain how viscosity, gas pressure, and the environment affect the eruptive style of a volcano.

Low-viscosity (basaltic) melts allow gas bubbles to effervesce gradually; high-viscosity (intermediate to silicic) melts trap gas internally. Further, these melts tend to become clogged within the volcanic chimney, causing gas pressure to build up ultimately to catastrophic levels. Higher levels of gas pressure produce more powerful explosions. The environment can affect the cooling rate of lava; for example, basaltic lava cannot flow as far from a submarine volcano as from one on land because it cools so quickly in water.

9. Describe the activity in the mantle that leads to hot-spot eruptions.

At the base of the mantle (core-mantle boundary) a body of mantle material may become so heated that it will rise up relative to cooler, denser surroundings. Contact with the lithosphere generates partial melting to produce basaltic magma, which rises up through the crust to form a volcano at the surface.

10. How do continental rift eruptions form flood basalts?

Hot plumes of mantle material that arise in areas of rifting meet a set of circumstances—decompression and the formation of fissures—that are conducive to volcanic output. Above a plume, relatively large amounts of partial melting produce massive quantities of basaltic magma, which rise through fissures and, due to low viscosity, spread outward to form uniform sheets (in the manner of aqueous floods).

11. How are black smokers formed, and where are they found?

Black smokers, found along mid-ocean ridges, arise because water traveling through the pores in these regions of high heat flow rapidly vents outward into the cooler ocean. Part of the dissolved load carried by these hot-water vents becomes insoluble in contact with the cooler ocean, bringing precipitation of dark crystals of sulfide minerals and giving the water gushing through the vents its smoky appearance.

12. What is the "Ring of Fire"?

The "Ring of Fire" refers to the volcanoes that are so abundant around the margins of the Pacific Ocean due to subduction and consequent formation of volcanic arcs.

13. Contrast an island volcanic arc with a continental volcanic arc.
   In an island volcanic arc, the overriding plate at the subduction zone is composed of oceanic lithosphere. Resultant volcanism produces an island chain, such as the Aleutians, dominated by basalt. In contrast, when the overriding plate is continental lithosphere, a chain of terrestrial volcanoes (termed a continental volcanic arc) extruding intermediate, andesitic lava is produced.

14. How are volcanoes sometimes beneficial?
   Volcanoes gave rise to the crust and oceans; they provide nutrients for soils and form islands in the middle of oceans that would otherwise be uninhabitable for terrestrial organisms such as ourselves.

15. List some of the major volcanic hazards.
   Lava flows, falling ash, pyroclastic flows, explosive eruption, landslides, lahars, earthquakes, tsunamis, poison gas.

16. How do scientists predict volcanic eruptions?
   Measured increase in heat flow, changes in the shape of volcanoes, and increasing incidence of earthquakes and gaseous emissions may signal that an eruption is imminent.

Test bank

1. In 79 C.E., the citizens of Pompeii in the ancient Roman Empire were buried by pyroclastic debris derived from an eruption of _____.
   A. Mt. Olympus
   B. Olympus Mons
   C. Mt. Vesuvius
   D. Mt. St. Helens

2. The smoky cloud that rises from the vent of an actively erupting volcano is composed of _____.
   A. smoke from wildfires inside the volcanic chimney
   B. smoke from wildfires on the outer slopes of the volcano that become funneled into the crater
   C. fine pyroclastic debris (ash) suspended in the air
   D. a continuous fountain of dark, basaltic lava

3. Most terrestrial volcanic glass _____.
   A. is highly silicic in composition
   B. is coarse grained
   C. consists of microscopic crystals
   D. freezes from lavas with low viscosity

4. Basaltic lavas _____.
   A. contain more iron and magnesium than rhyolitic lavas
   B. contain more silica than rhyolitic lavas
   C. are more viscous than rhyolitic lavas
   D. contain a greater proportion of trapped volatiles than rhyolitic lavas

5. Pahoehoe _____.
    A. forms when basaltic lava flows cease flowing and solidify simultaneously
    B. has a smoother texture than aa
    C. is easier to walk on than aa is
    D. All of the above are correct.

6. Basaltic lavas which solidify at the surface before flow ceases fracture irregularly, producing a sharp-surfaced lava rock named _____.
    A. pahoehoe   B. aa
    C. pumice   D. hyaloclasite

7. Pillow lavas are associated with _____.
    A. continental rhyolitic eruptions   B. continental basaltic eruptions
    C. submarine rhyolitic eruptions   D. submarine basaltic eruptions

8. Rhyolitic lavas _____.
    A. do not flow as far from the vent as basaltic lavas do
    B. cool much more slowly than basaltic lavas do
    C. are associated with volcanoes that almost never emit pyroclastic debris
    D. All of the above are correct.

9. Ash, cinders, and blocks are all types of _____.
    A. pyroclastic debris   B. lava flows
    C. Pele's hair   D. volcanoes

10. The difference between tephra and tuff is that _____.
    A. tephra is created in ash falls, whereas tuff is created in pyroclastic flows
    B. tephra is unlithified, whereas tuff is lithified
    C. tephra is always silicic, whereas tuff is always basaltic
    D. All of the above are correct.

11. The lithification of material from a pyroclastic flow forms a rock called _____.
    A. metabasalt   B. ignimbrite
    C. migmatite   D. tuff

12. Volcanic bombs are _____.
    A. pyroclastic blocks that acquire aerodynamic shapes during flight out of the volcanic vent
    B. explosive bodies of lava with high volatile content
    C. cinders that explode upon impact with the ground
    D. used by geologists to set off small eruptions in volcanoes that are deemed potentially dangerous

13. Pele's hair _____.
    A. consists of thin strands of basaltic pyroclastic debris
    B. acquires aerodynamic torpedo-like shapes during flight out of the volcanic vent

C. forms only at the leading edge of basaltic lava flows
D. was instrumental in scoring the winning goal to defeat Uruguay in the 1958 World Cup finals

14. In 1902, a famous, deadly pyroclastic flow killed thousands of people on the Caribbean island of _____.
    A. Puerto Rico   B. Martinique
    C. Aruba   D. Jamaica

15. A fast-moving flow consisting of a mixture of water and pyroclastic debris is termed a _____.
    A. lahar   B. glowing avalanche
    C. flood basalt   D. stratovolcano

16. Gases that are abundantly emitted by volcanoes include _____.
    A. water vapor, carbon dioxide, and sulfur dioxide
    B. oxygen, ozone, and water vapor
    C. oxygen, hydrogen, and neon
    D. carbon dioxide, carbon monoxide, and oxygen

17. The characteristic "rotten egg" smell of many active volcanoes is derived from _____.
    A. iron and magnesium within lava
    B. carbon dioxide gas
    C. stale water within the magma chamber
    D. hydrogen sulfide gas

18. Explosive or voluminous eruptions may cause the volcano to collapse upon the floor of the (now empty) magma chamber, producing a broad depression termed a _____.
    A. crater   B. lahar   C. caldera   D. fissure

19. Of the three primary forms of subaerial volcanoes, _____ are the largest in areal extent.
    A. stratovolcanoes   B. cinder cones   C. shield volcanoes

20. Of the three primary forms of subaerial volcanoes, _____ have the most gently sloping sides, due to the low viscosity of the basaltic lavas which form them.
    A. stratovolcanoes   B. cinder cones   C. shield volcanoes

21. Of the three primary forms of subaerial volcanoes, _____ consist of a simple, conical pile of tephra.
    A. stratovolcanoes   B. cinder cones   C. shield volcanoes

22. Of the three primary forms of subaerial volcanoes, _____ consist of alternating layers of tephra and solidified lava.
    A. stratovolcanoes   B. cinder cones   C. shield volcanoes

23. Of the three primary forms of subaerial volcanoes, _____ are sometimes referred to as "composite volcanoes."
    A. stratovolcanoes   B. cinder cones   C. shield volcanoes

24. Mt. Fuji in Japan is an example of a _____.
    A. stratovolcano   B. cinder cone   C. shield volcano

25. Olympus Mons, the largest known volcano in the Universe, is an example of a _____.
    A. stratovolcano   B. cinder cone   C. shield volcano

26. Olympus Mons, the largest known volcano in the Universe, is found on _____.
    A. Earth   B. Mars
    C. Neptune   D. Io, a moon of Jupiter

27. All volcanic eruptions pass through the crater at the volcanic summit.
    A. true   B. false

28. Nonviolent eruptions characterized by extensive flows of basaltic lava are termed _____.
    A. pyroclastic   B. effusive   C. explosive

29. Whether an eruption will primarily produce lava flows or pyroclastic debris is influenced by the _____.
    A. viscosity of the lava
    B. composition of the lava
    C. proportion of volatiles within the lava
    D. All of the above are correct.

30. As compared to subaerial basaltic lavas, submarine basaltic lavas differ in that they _____.
    A. always produce violent pyroclastic debris flows
    B. produce large crystals of pyroxene and plagioclase
    C. form pillow-like mounds because they cannot flow as far from their source
    D. All of the above are correct.

31. Phreatomagmatic eruptions take place when _____.
    A. volatiles effervesce prior to lava flow
    B. water enters the magma chamber and forms steam
    C. basaltic lava clogs the chimney
    D. lava or pyroclastic debris erupts by bursting through the sides of the volcano

32. Phreatomagmatic eruptions are _____.
    A. explosive (pyroclastic)   B. effusive (dominated by lava flows)

33. Hot-spot volcanoes _____.
    A. can arise from the ocean floor
    B. can arise on continents
    C. may arise in the interior of lithospheric plates
    D. All of the above are possibilities.

34. The hot-spot track associated with the Hawaiian Islands and Emperor Seamounts _____.
    A. shows that the Pacific Plate has been stationary over the last 30 million years
    B. occurs along a divergent plate boundary
    C. occurs along a convergent plate boundary
    D. shows that the Pacific Plate has been moving northwest for the last 30 million years

35. At continental rifts, vast bodies of basaltic lava flow forth from fissures, forming _____.
    A. flood basalts   B. pyroclastic flows   C. ash-fall tuffs

36. Iceland rises above the Atlantic Ocean as a result of _____.
    A. normal mid-ocean ridge activity
    B. a submarine hot-spot located within the interior of a plate
    C. a submarine hot-spot located along a mid-ocean ridge
    D. subduction of an oceanic plate underneath the continental Eurasian plate

37. Eruptions emitted through elongate cracks at the surface, as opposed to through the vents of volcanic cones, are termed _____.
    A. hot-spot eruptions   B. fissure eruptions
    C. phreatomagmatic eruptions   D. pyroclastic eruptions

38. The new island of Surtsey near Iceland was initially a _____ but is now a _____.
    A. cinder cone; stratovolcano
    B. stratovolcano; shield volcano
    C. shield volcano; cinder cone
    D. subaerial volcano; submarine volcano

39. Of the three major types of volcanoes, _____ erode away fastest after the cessation of eruptions.
    A. stratovolcanoes   B. cinder cones   C. shield volcanoes

40. Volcanoes produce no other hazards besides lava flows and pyroclastic debris.
    A. true   B. false

Chapter 10
A Violent Pulse: Earthquakes

Learning objectives

1. Earthquakes are episodes of shaking within the Earth. Most, though not all, earthquakes represent shock arising from displacement along a fault.

2. Strike-slip faults occur along nearly vertical planes; opposite sides are horizontally offset. Oblique fault planes give rise to a hanging wall and footwall (with the hanging wall vertically above the footwall at any point along the fault). If the foot wall slides up with respect to the hanging wall, the fault is termed normal. If the hanging wall slides up with respect to the footwall, the fault is termed reverse (or thrust, if the fault dips shallowly).

3. Students should know the four types of seismic waves, which two types occur at the surface, which two travel through the interior, and which travels fastest. They should be able to distinguish the difference between the compressional motion of the P-wave and the shear motion of the S-wave. They should know that earthquakes travel fastest through solid igneous rock, slower through sedimentary rock, slower yet through sediment, and slowest through liquids (S-waves will not pass through a liquid).

4. They should know the principal components of a classic seismograph, and how travel-time curves can be used from three stations to isolate the epicenter of an earthquake.

5. Distinctions among the Mercalli, Richter, and seismic moment scales should be understood.

6. The coincident alignment between most belts of seismic activity and plate boundaries should be noted, with the understanding that intermediate- and deep-focus earthquakes occur only at convergent margins (plate boundaries).

7. Earthquakes induce a variety of damaging hazards including ruptured gas lines, fires, tsunamis, landslides, and sediment liquefaction.

Summary from the text

Earthquakes are episodes of ground shaking, caused when earthquake waves reach the ground surface. Earthquake activity is called seismicity.

Most earthquakes happen when rock breaks during faulting. A fault is a fracture on which sliding occurs. The place where rock breaks and earthquake energy is released is called the focus, and the point on the ground directly above the focus is the epicenter.

We can distinguish between normal, reverse, thrust, and strike-slip faults on the basis of the relative motion of rock across the fault. The amount of movement is called the displacement.

Active faults are faults on which movement is currently taking place. Inactive faults ceased being active long ago but still can be recognized because of the displacement across them. Displacement on active faults that intersect the ground surface may yield a fault scarp. The intersection of the fault with the ground is the fault trace. Faults that do not intersect the ground are blind faults.

During fault formation, rock elastically strains, then cracks form. Eventually, the cracks link to form a through-going rupture on which sliding occurs. When this happens, the elastically strained rock breaks and vibrates, and this generates an earthquake.

Slip may occur on a fault more than once. Friction resists sliding until stress acting on the fault gets large enough. Thus faults exhibit stick-slip behavior, in that they move in sudden increments.

Earthquake faulting is a type of brittle deformation. Earthquakes in the continental crust can only happen in the brittle, upper part of the crust. At depth, where rocks become ductile, earthquakes don't occur.

Earthquake energy travels in the form of seismic waves. Body waves, which pass through the interior of the Earth, include P-waves (compressional waves) and S-waves (shear waves). Surface waves, which pass along the surface of the Earth, include Rayleigh waves and Love waves.

We can detect earthquake waves using a seismograph. A weight inside this instrument stays fixed in position, while a pen attached to the weight traces out seismic waves on a paper cylinder attached to the frame.

Seismograms demonstrate that different earthquake waves arrive at different times because they travel at different velocities. Using the difference between P-wave and S-wave arrival times, graphed on a travel-time curve, seismologists can determine the distance from a seismograph station to an earthquake epicenter and can then pinpoint the epicenter location.

The Mercalli intensity scale measures damage caused by an earthquake. The Richter magnitude scale measures the size of the largest recorded earthquake wave on a seismogram. The seismic-moment magnitude scale takes into account the amount of slip, the length and depth of the rupture, and the strength of the ruptured rock.

The largest earthquakes have a Richter magnitude of about 8.9. A magnitude 8 earthquake yields about ten times as much ground motion as a magnitude 7 earthquake and releases about 33 times as much energy. Great earthquakes have a magnitude greater than about 8.0.

Most earthquakes occur in seismic belts, or zones, of which the majority lie along plate boundaries. Intraplate earthquakes, which happen in the interior of plates, are relatively infrequent but can be as large as the one that occurred in New Madrid, Missouri.

Different kinds of earthquakes happen at different kinds of plate boundaries. Shallow-focus earthquakes associated with normal faults occur at divergent plate boundaries and in rifts. Earthquakes associated with thrust and reverse faulting occur at convergent and collisional boundaries. At convergent plate boundaries, we also observe intermediate- and deep-focus earthquakes, which define the Wadati-Benioff zone. Shallow-focus strike-slip earthquakes occur along transform boundaries.

Earthquakes can be induced at some locations by injecting water into the ground.

Earthquake damage results from ground shaking (which can topple buildings), landslides (set loose by vibration), sediment liquefaction (the transformation of compacted clay into a muddy slurry), fire, and tsunamis (giant waves).

Seismologists can predict that earthquakes are more likely in seismic zones than elsewhere and can determine the recurrence interval for great earthquakes (the average time between successive events) by studying geologic features such as the distribution of sand volcanoes and disrupted bedding in a sequence of recent strata. But it may never be possible to pinpoint the exact time and place at which an earthquake will take place.

Earthquakes may be more likely at seismic gaps (places where no recent earthquakes have happened) along seismic belts, presumably because stress is building in these locations.

Earthquake hazards can be reduced with better construction practices and zoning and by knowing what to do during an earthquake.

Answers to review questions

1. Compare normal, reverse, and strike-slip faults.

Normal and reverse faults occur on planes that are oriented at some angle between horizontal and vertical. In cross-sectional view, drawing a vertical line intersecting the fault plane illustrates the distinction between a hanging wall (block of material on one side of the fault, including the rock above the intersection of the fault plane and the vertical line) and a footwall (material on the opposite side of the fault, including rock below the intersection of the fault and the vertical line). In a normal fault, the hanging wall has slid downward with respect to the footwall (mnemonic phrase: "a hanging wall will normally fall"). In a reverse fault, the hanging wall has slid upward with respect to the footwall.

Strike-slip faults occur on planes that are vertical or nearly so. Offset consists of the blocks sliding past one another laterally, along the line of strike (intersection between the fault plane and ground surface).

2. Why are blind faults even more worrisome than faults with known surface breaks?

Blind faults can produce seismic hazards in areas where earthquakes are not expected (blind faults may go undetected until they produce an earthquake).

3. What is the difference between stress and strain? What are the three kinds of stress?

Stress is the compression, tension, or shear that develops as the result of force (numerically measured as force per unit area). The three kinds of stress are compressive (in which material is squeezed together), tensile (or dilatational, in which material is pulled apart), and shear (in which one end of the material is forced laterally with respect to an opposite end). Strain is a change in shape that occurs as a response to stress.

4. Describe elastic strain.

An elastic strain is one in which the strained object will resume its prestressed shape after the stress is removed.

5. Compare brittle and ductile deformation.

Brittle deformation is the development of fractures (as when two kids pull on a piece of peanut brittle); ductile deformation occurs when material bends and flows without breaking (as when two kids pull on a piece of taffy).

6. Describe the motions of the four types of seismic waves. Which are body waves, and which are surface waves? What are their relative velocities?

    P-waves and S-waves are body waves, traveling through the Earth's interior. P-waves are compressional waves, meaning that material affected by the waves oscillates back and forth in the direction of wave propagation, alternately compressing and expanding like an accordion. S-waves are shear waves, in which the material affected by the wave oscillates perpendicularly to the direction of wave propagation.

    L- (Love) and R- (Rayleigh) waves are surface waves. L-waves are shear waves that propagate at the surface, producing lateral oscillations perpendicular to the direction of propagation. Motion in R-waves has both shear and compressional components, similar to propagating waves in the deep ocean; material affected by the wave takes an elliptical path, up and down (perpendicular to propagation) and back and forth (in the direction of propagation).

    P-waves travel fastest, S-waves are considerably slower, and surface waves are slower yet.

7. How do the velocities of P-waves change as they pass through different materials? Which materials have the highest seismic velocity? Which have the slowest?

    P-waves travel most rapidly (8 km/s) through the ultramafic rock of the mantle. Within the crust, P-wave relative velocities through the following media are ordered:

basalt > granite > sedimentary rock > loose sediment > water

8. Explain how the vertical and horizontal components of an earthquake are detected on a seismograph.

    Vertical motions can be detected with a heavy weight suspended downward by a spring from a sturdy frame bolted into the ground, with a pen extending laterally and in contact with a rotating paper cylinder that is oriented vertically. Horizontal motion detectors have the weight attached to a bar that is free to rotate horizontally, but not vertically, and a pen in contact with a horizontally oriented paper cylinder.

9. Why was the worldwide seismic network of military as well as scientific value?

    They record vibrations set off by underground nuclear explosions.

10. Compare the Mercalli, Richter, and seismic-moment scales in terms of what they measure.

    The Mercalli intensity scale is a relative scale (I–XII) depicting a subjective assessment of the damage inflicted by an earthquake upon a given city, village, or settlement. The Richter scale is the magnitude of the largest seismic oscillation as recorded by a seismograph that is situated 100 km away from the epicenter, as measured directly on a seismogram from this ideally situated seismograph station. The seismic-moment scale is a product of the amount of slip incurred along the fault during the earthquake, the length and depth of the rupture, and the strength of the fractured rock.

11. How does seismicity on mid-ocean ridges compare with seismicity at convergent or transform boundaries?

At mid-ocean ridges, earthquakes associated with normal faults occur along the ridge segments (strike-slip faulting occurs along the transform faults that connect them). All mid-ocean ridge earthquakes have shallow foci. At transform plate boundaries, strike-slip faulting predominates, with uniformly shallow foci as well.

At convergent boundaries (subduction zones), a variety of earthquakes occur. The subducting plate may be offset by both normal and thrust (reverse) faults, with the largest thrust faults arising at the contact plane between plates. The overriding plate develops reverse faults as well. The subducting slab may continue to generate earthquakes at depths considered intermediate and deep (up to 670 km).

12. What is the Wadati-Benioff zone, and why was it important in understanding plate tectonics?

The Wadati-Benioff zone is an oblique band of earthquake foci that descend into the mantle along with the subducted lithospheric slab, extending down to a depth of 670 km. Because the mantle rocks at this depth are too hot to experience earthquakes, its existence proves that the surficial lithosphere is indeed subducted down into the Earth's interior, clearing the way for new lithosphere to be produced to either side of mid-ocean ridges.

13. Describe the different types of damage caused by earthquakes.

The shaking of the ground due to surface waves can destroy buildings, bridges, and other constructions. Seismically induced water waves (seiching in lakes and tsunamis in the oceans) can destroy nearshore habitations. Earthquakes can trigger landslides, and earthquake damage may be locally aggravated if liquefaction of clay-rich soils occurs (seismic waves can cause coordinated water molecules within clay minerals to shake loose, forming a water and clay mixture that behaves as a liquid).
Earthquakes can rupture gas lines to produce catastrophic fires and may cause unsanitary conditions that harbor disease.

14. What is a seiche, and how does it move?

A seiche is a seismically induced span of abnormally large waves traveling back and forth at the surface of a lake or other still body of water.

15. Explain how liquefaction occurs in an earthquake and how it can cause damage.

Seismic waves can cause coordinated water and mica-like molecules within clay minerals to shake loose, forming a water and clay mixture that behaves as a liquid. Buildings with their foundations in clay that has been subjected to liquefaction may topple over or slide down if built on a slope.

16. How are long-term and short-term earthquake predictions made?

Long-term earthquake predictions are on the basis of recorded earthquake occurrences. Within a designated time span, higher probabilities of future earthquakes are assigned to regions that have historically been seismically active, and to smaller earthquakes as compared to large earthquakes at any one particular region. More specifically, within belts of seismicity, seismic gaps (sections within seismically active belts that have not experienced as many recent earthquakes) may be "due" for an earthquake.

Short-term predictions are generally unreliable, but swarms of small earthquakes and sudden deformation of rock (perhaps evidenced by a change in the water table) may forebode of an earthquake to come.

17. Why is it difficult to make accurate short-term predictions?

Seismic activity may behave somewhat randomly or chaotically (so strongly sensitive to initial conditions, involving a multitude of variables, so as to be effectively random).

18. What types of structure are most prone to collapse in an earthquake? What types are most resistant?

Unreinforced brick and concrete buildings are highly prone to collapse. Wood- and steel-framed buildings and reinforced concrete buildings fare much better.

19. What should you do when you feel an earthquake starting?

If you're outside, stay there. If inside, get under a heavy table or a door frame near the center of the edifice.

Test bank

1. Geologists who specifically study earthquakes are called _____.
   A. seismologists
   B. paleontologists
   C. vulcanologists
   D. speleologists

2. All earthquakes are produced when previously cohesive bodies of rock are split along a sliding surface.
   A. true   B. false

3. A surface along which rock on opposed sides is offset by sliding during an earthquake is called a _____.
   A. joint   B. fault   C. fold   D. wall

4. At any point along the surface of an oblique (nonvertical) fault, the _____.
   A. hanging wall lies vertically above the footwall
   B. footwall lies vertically above the hanging wall
   C. hanging wall lies to the left of the footwall
   D. footwall lies to the left of the hanging wall

5. If, during an earthquake, a hanging wall slides upward relative to a footwall, the fault is termed _____ if the fault is steep (closer to vertical than horizontal).
    A. normal   B. reverse   C. strike-slip   D. thrust

6. If, during an earthquake, a hanging wall slides upward relative to a footwall, the fault is termed _____ if the fault is shallow (much closer to horizontal than vertical).
    A. normal   B. reverse   C. strike-slip   D. thrust

7. If, during an earthquake, a footwall slides upward relative to a hanging wall, the fault is termed _____.
    A. normal   B. reverse   C. strike-slip   D. thrust

8. If a fault is nearly vertical in orientation and the two walls of rock on opposite sides slide past one another horizontally, the fault is termed _____.
    A. normal   B. reverse   C. strike-slip   D. thrust

9. The quantity of offset that occurs along a fault is termed _____.
    A. fault gouge   B. the fault gauge
    C. displacement   D. accumulation

10. All discovered faults are likely to experience earthquakes in the next few hundred years.
    A. true   B. false

11. The intersection between a fault plane and the ground surface is called the _____.
    A. dip line   B. plunge
    C. fault trace   D. seismic interface

12. Which type of fault does not, by definition, have a fault trace?
    A. normal fault
    B. reverse fault
    C. blind fault
    D. None of the above; all faults have recognizable traces.

13. Which type of stress is exerted by a pair of scissors cutting into a piece of paper?
    A. compressive stress   B. tensile stress
    C. shear stress   D. emotional stress

14. Which type of stress confronts someone who dives deep into the ocean?
    A. compressive stress   B. tensile stress
    C. shear stress   D. ellipsoidal stress

15. Which type of stress is exhibited when an object is pulled apart in multiple directions?
    A. compressive stress   B. tensile stress
    C. shear stress   D. emotional stress

16. The relationship between stress and strain is that _____.
    A. stress is a result of strain
    B. strain is a result of stress
    C. "stress" and "strain" are exact synonyms
    D. "strain" is used in everyday speech to connote weariness but has no meaning in geology

17. A primary force opposing motion on all faults is _____.
    A. magnetic attraction among iron-rich minerals
    B. gravity
    C. friction
    D. Van der Waal's force

18. Periods of intermittent sliding on a fault as a result of the release of stress during episodes of displacement, followed by stress buildup to the point that the fault is reactivated, is termed _____.
    A. chaotic faulting   B. thrust faulting
    C. stick-slip behavior   D. reverse faulting

19. Aftershocks following a major earthquake _____.
    A. may continue for days after the initial earthquake
    B. are mostly much smaller than the original earthquake
    C. may occur on the same fault as the original earthquake, or a different fault
    D. All of the above are correct.

20. Faulting and earthquakes are examples of _____.
    A. brittle behavior   B. ductile behavior

21. Earthquake waves that pass through the interior of the Earth are termed _____.
    A. interior waves   B. Rayleigh waves
    C. surface waves   D. body waves

22. Generally which type of earthquake waves travel fastest?
    A. interior waves   B. Rayleigh waves
    C. surface waves   D. body waves

23. Surface waves _____.
    A. travel more rapidly than body waves
    B. produce most of the damage to buildings during earthquakes
    C. are the first waves initially produced in an earthquake
    D. are the first waves to arrive at a seismograph station after an earthquake

24. Body waves include _____.
    A. both S- and P-waves   B. both L- and R-waves
    C. both surface and interior waves   D. P-waves only

25. Which type of earthquake wave travels fastest?
    A. L-wave	B. P-wave	C. R-wave	D. S-wave

26. Earthquake waves propagate most rapidly through _____.
    A. sediment	B. sedimentary rock
    C. igneous rock	D. water

27. Vertical motion seismographs record earthquakes through the production of a squiggly diagram called a _____.
    A. wave sheet	B. seismogram
    C. pictogram	D. camera lucida

28. The worldwide seismic network has played an important role in human political history because seismographs detect not only the waves emitted by earthquakes but also _____.
    A. the vibrations of submarines in deep water
    B. underground nuclear explosions
    C. the activities of astronauts on the Moon
    D. the flight paths of Scud missiles

29. The point within the Earth where an earthquake takes place is termed the _____.
    A. focus	B. epicenter
    C. eye of the fault	D. vertex

30. The point on Earth's surface directly above the point where an earthquake occurs is termed the _____.
    A. focus	B. epicenter
    C. eye of the fault	D. vertex

31. How many seismic stations are necessary to find the epicenter of an earthquake?
    A. one	B. two	C. three	D. four

32. Which earthquake intensity scale assesses the effects of an earthquake upon manmade structures?
    A. Richter scale
    B. Mercalli scale
    C. seismic-moment magnitude scale

33. Which earthquake intensity scale is most commonly reported by news media following a major earthquake?
    A. Richter scale
    B. Mercalli scale
    C. seismic-moment magnitude scale

34. Which earthquake intensity scale takes into account the type of rock that has been fractured?
    A. Richter scale
    B. Mercalli scale
    C. seismic-moment magnitude scale

35. Which earthquake intensity scale measures the amplitude of deflection of a seismograph pen, standardized to a idealized distance of 100 km between epicenter and seismograph?
    A. Richter scale
    B. Mercalli scale
    C. seismic-moment magnitude scale

36. Which earthquake intensity scale varies from locality to locality for a single earthquake?
    A. Richter scale
    B. Mercalli scale
    C. seismic-moment magnitude scale

37. Earthquakes are likely to occur along _____.
    A. convergent plate boundaries only
    B. divergent plate boundaries only
    C. transform plate boundaries only
    D. all three major types of plate boundaries

38. Medium- and deep-focus earthquakes occur along _____.
    A. convergent plate boundaries only
    B. divergent plate boundaries only
    C. transform plate boundaries only
    D. all three major types of plate boundaries

39. Earthquakes that occur in a band called the _____ can be used to track the motion of subducted oceanic lithosphere.
    A. Wegener belt
    B. seismic gap
    C. Wadati-Benioff zone

40. Virtually all of the deaths attributed to major earthquakes have resulted from the collapse of buildings.
    A. true   B. false

Chapter 11
Crags, Cracks, and Crumples: Crustal Deformation and Mountain Building

Learning objectives

1. Students should understand normal, reverse, thrust, detachment, right-lateral and left-lateral faults, including the orientation of the fault plane and sense of motion in each case. They should know how faults and joints differ, and the evidence used in the field to diagnose the presence of a fault.
2. They should be able to recognize the four major folds (synclines, anticlines, domes, and basins) in both map view and cross-section.
3. They should be able to distinguish between brittle and ductile behavior, recognizing that faults exemplify the former and folds the latter. They should know that heat, pressure, and gradual application of stress favor ductile behavior in rocks.
4. Orogeny leads to the production of all three rock types through the upward migration of plutons, regional and contact metamorphism, and erosion. Mountain ranges are common along convergent and divergent plate boundaries, at collision zones and continental rifts.
5. Isostatic equilibrium denotes the balance between the weight of mountain ranges and the buoyant support they receive from the more dense mantle below. Mountain ranges are underlain by thick crustal roots, just as most of an ice cube floats below the surface of the water. As mountain ranges erode, their crustal roots are pressed upward (an example of isostatic compensation).
6. The familiar montane topography is the result of erosion. Once the rate of uplift falls behind the rate of erosion, the mountain range will begin to wear flat, a process that generally takes tens of millions of years.
7. Flat, low-lying regions that have not been exposed to orogenic deformation for more than one billion years are termed cratons. In the central portion of the craton, termed the shield, metamorphic rocks are exposed at the surface. The region on the flanks of the craton, where the metamorphics are overlain by sediments, is termed the platform.

Answers to review questions

1. What are the changes that rocks undergo in an orogenic belt like the Alps?
    In orogenic belts, rock undergoes deformation as a response to stress. Deformation can include faulting, jointing, folding, and the development of metamorphic foliation.

2. What is the difference between brittle and ductile deformation?
    See Chapter 10, question 5.

3. What factors influence whether a rock will behave in brittle or ductile fashion?
    Temperature: hot rocks are more ductile than cool rocks.
    Pressure: rocks under very high pressure behave more ductilely than those at low pressure.

Deformation rate: sudden changes, such as onset of tensile stress, are more likely to produce brittle behavior than are gradual changes.

Rock type: some rocks, such as halite, have a proclivity to behave ductilely.

4. How are stress and strain different?
See Chapter 10, question 3.

5. How is a fault different from a joint?
A fault is a fracture along which there has been displacement; a joint is a fracture without displacement.

6. Compare the motion of normal, reverse, and strike-slip faults.
See Chapter 10, question 1.

7. How do you recognize faults in the field?
Offset of layers on opposite sides of the fault, the development of drag folds along the fault interface, shattered rock (fault breccia), powdered rock (fault gouge), and slickensides (polished fault surfaces) are all clues used to identify faults. (Note: Slickensides are smooth to the touch in the direction of fault motion but very rough if brushed with the fingers in the opposite direction.)

8. Describe the differences between an anticline, a syncline, and a monocline.
Anticlines (unless overturned) are convex-upward arches. Synclines (unless overturned) are concave-upward troughs. Monoclines are step-like folds.

9. Explain how certain kinds of igneous, sedimentary, and metamorphic rocks are formed during orogeny.
Extrusive and intrusive rocks are formed at convergent margins, in association with melting at a subduction zone. Uplift of rocks leads to erosion and the development of vast quantities of sediment, which are transported from the mountain face to alluvial fan deposits. Sediment in the fans ultimately lithifies to form sedimentary rock. Thermal (contact) metamorphism occurs adjacent to rising plutons within mountain ranges, and thrust-fault loading at collision zones leads to deep burial and dynamothermal (regional) metamorphism, producing foliated metamorphic rocks.

10. Describe the principle of isostasy.
The tendency of gravity to pull down a body of lithosphere (such as a mountain range) is counterbalanced by the buoyant support of denser material below the body.

11. What happens to the isostatic equilibrium of a mountain range as it is eroded away?
The submerged base of the mountain range (its crustal root) will rise upward in response to the alleviation of stress, to form a new isostatic equilibrium.

12. What happens to a mountain range when its rate of uplift slows down?
    Its crustal root will continue to grow so long as the rate of uplift exceeds the rate of erosion. Once erosion becomes the more rapid process, the range will begin to be buoyed upward by the asthenosphere.

13. Discuss the processes by which mountain belts are formed in convergent margins, in continental collisions, and in continental rifts.
    At convergent margins, if the overriding plate is a continent, a continental volcanic arc is formed. Collision may occur between the overriding continental plate and an exotic terrane (island arc or microcontinent), which will not subduct, but rather merges into the continent, elevating the continental volcanic arc.
    At collision zones between continents, massive mountain ranges are produced. Thrust faulting at a fold-thrust belt loads the material from one continent above the other, leading to deep burial of rocks and regional metamorphism.
    At continental rifts, the lithosphere is thinned by stretching, and the hot asthenosphere rises to compensate, heating the thinned lithosphere above and thereby making it less dense. Its reduced density causes the thin lithosphere to rise up to reestablish isostatic equilibrium. Normal faulting produces regions of relative uplift, termed fault-block mountains, to the sides of rift basins.

14. How are the structures of a craton different from a typical orogenic belt?
    At a shield within a craton, very ancient (greater than one billion years old), highly metamorphosed rocks are exposed at the surface. These metamorphic rocks formed in ancient orogenies and have been eroded flat. These exposed rocks were once deep within the core of a mountain range, but cycles of erosion and isostatic uplift have worn away all other evidence of ancient mountains.

Test bank

1. Mt. Everest, the tallest mountain in the world, is located on the continent of _____.
    A. Africa	B. Asia	C. North America	D. Europe

2. An episode of mountain building is termed a(n) _____.
    A. orogeny	B. phylogeny	C. aureole	D. slickenside

3. After uplift is completed, a large mountain range should be eroded flat within _____.
    A. tens of thousands of years	B. hundreds of thousands of years
    C. tens of millions of years	D. hundreds of millions of years

4. Nearly all of the present mountain ranges are the products of single orogenic events.
    A. true	B. false

5. Deformation brought on by orogeny can _____.
   A. metamorphose rock
   B. produce folds in rock
   C. produce faulting in rock
   D. All of the above are produced by orogeny.

6. Change in shape, induced by stress, is termed _____.
   A. plastic deformation   B. pressure release
   C. strain   D. metamorphosis

7. A body of rock affected by tensile stress will likely undergo _____.
   A. shortening   B. stretching   C. shear strain

8. A body of rock affected by compressive stress will likely undergo _____.
   A. shortening   B. stretching   C. shear strain

9. A hot body of rock is more likely to exhibit _____ than is a cold body of rock.
   A. brittle behavior   B. ductile behavior

10. A body of rock under high pressure is more likely to exhibit _____ than is a body of rock at low pressure.
    A. brittle behavior   B. ductile behavior

11. A body of rock to which a sudden, rapid stress has been applied is more likely to exhibit _____ than is a body of rock subjected to a gradually applied stress.
    A. brittle behavior   B. ductile behavior

12. Earthquakes only occur _____ the brittle/ductile transition depth of 10–15 km.
    A. above   B. below   C. at or near

13. Force per unit area is termed _____.
    A. stress   B. strain   C. power   D. work

14. The distinction between joints and faults is that _____.
    A. faults are joints that are greater than one square meter in areal extent
    B. faults are fractures along which displacement has occurred; displacement does not occur along joints
    C. joints are fractures along which displacement has occurred; displacement does not occur along faults
    D. There is no distinction; the two terms are synonymous.

15. A joint always occurs as a single, isolated plane within a rock.
    A. true   B. false

16. Most fault surfaces, like joints, are roughly planar in orientation.
    A. true   B. false

17. Motion along all faults is either strike-slip or dip-slip; combinations of these two types of displacement are never found together in a single fault.
    A. true   B. false

18. Normal, reverse, and thrust are all examples of _____ faults.
    A. strike-slip   B. dip-slip   C. oblique-slip

19. Right-lateral and left-lateral are both examples of _____ faults.
    A. strike-slip   B. dip-slip   C. oblique-slip

20. In the above map the vertical, north-south trending fault is a _____ fault.
    A. normal dip-slip   B. reverse dip-slip
    C. right-lateral strike-slip   D. left-lateral strike-slip

21. It is possible for offset along an oblique-slip fault to have both _____ components.
        A. normal and reverse
        B. right-lateral and left-lateral
        C. normal and left-lateral

22. Movement along faults often produces sharply angled rock fragments termed _____.
        A. fault gouge   B. rock flour
        C. fault breccia   D. slickensides

23. Shear stress at sufficient depth within a fault plane can induce ductile shear, forming a fine-grained metamorphic rock named _____.
    A. ignimbrite   B. gneiss   C. mylonite   D. migmatite

24. A basin bounded by normal faults, which dip toward the center of the basin, is termed a(n) _____.
    A. foreland basin   B. intracratonic basin
    C. horst   D. graben

25. Normal faults assume a more shallow dip angle with depth; when the fault plane becomes nearly horizontal, these faults are termed _____.
    A. thrusts   B. folds
    C. detachments   D. decollements

26. A fold shaped like an elongate arch is a(n) _____.
    A. anticline   B. basin   C. dome   D. syncline

27. A fold shaped like an elongate trough is a(n) _____.
    A. anticline   B. basin   C. dome   D. syncline

28. A fold shaped like an upside-down bowl is a(n) _____.
    A. anticline   B. basin   C. dome   D. syncline

29. A fold shaped like an right-side-up bowl is a(n) _____.
    A. anticline   B. basin   C. dome   D. syncline

30. On a geologic map, an anticline appears as a series of parallel stripes with the _____ unit in the center.
    A. youngest   B. oldest

31. The central portion of high curvature on a fold is termed the fold _____.
    A. limb   B. hinge   C. midsection   D. thorax

32. Tectonic foliation, such as elongation of quartz grains, always occurs parallel to the original bedding plane of a body of rock.
    A. true   B. false

33. Orogenesis (mountain building) leads to the production of _____.
    A. metamorphic rocks only   B. igneous rocks only
    C. sedimentary rocks only   D. all three major rock types

34. Theoretically, there is no reason why mountains significantly taller than Mt. Everest might not one day arise on Earth.
    A. true   B. false

35. Continental crust is typically 35 km thick but may be up to _____ thicker under mountain ranges.
    A. 20%   B. 50%   C. 100%   D. 200%

36. The balance between the weight of a mountain range and the buoyancy provided by the underlying mantle is termed _____.
    A. punctuated equilibrium   B. homeostatic equilibrium
    C. isostatic equilibrium   D. osmotic equilibrium

37. Valleys and hillsides carved by glaciers are generally more _____ in comparison to those produced by rivers and streams.
    A. steep sided   B. shallow sided

38. Regions of continents that have not been subjected to orogeny during the past one billion years are termed _____.
    A. exotic terranes
    B. accreted terranes
    C. cratons

39. The outer portion of a craton, where deformed rocks are covered by sediments, is termed the _____.
    A. shield   B. platform   C. convergent margin

40. Regions where Precambrian metamorphic rocks are exposed at the surface are termed _____.
    A. shields   B. platforms   C. convergent margins

Chapter 12
Deep Time: How Old Is Old?

Learning objectives

1. Students should become familiar with the geologic time scale. The ordered sequence of eons (Priscoan, Archean, Proterozoic, Phanerozoic) and of eras within the Phanerozoic eon, as well as starting dates for the eons and for the Mesozoic and Cenozoic eras.

2. They should be familiar with the distinction between relative (ordinal) time represented by the names of eras, periods, etc., and numerical time represented by dates measured in years. The source of most names derives from the fossil content of sedimentary rocks; numerical ages for boundaries on the time scale require radiometric dates for igneous (and some metaigneous) rocks combined with the observation of spatial relationships between these rocks and sedimentary strata.

3. Similarly, they should be able to understand and apply the principles that can be used to interpret the order of events that occurred to produce a given sequence of rocks (Steno's original continuity, horizontality, and superposition, as well as cross-cutting relationships, metamorphosed [baked] contacts, and inclusions).

4. Uniformitarianism ("the present is the key to the past") was an important step forward in geological thinking. Observing modern rates of deposition, erosion, and volcanic extrusion provides a first approximation for understanding ancient Earth conditions and further implies that the Earth must be very ancient (given the complexities of many outcrops, the past assemblies of supercontinents, etc.).

5. Fossil succession allows relative ages to be produced for many fossiliferous sedimentary rocks regardless of their spatial contexts. This allows for temporal correlation of widely separated strata on the basis of fossil content. Students should be able to produce overlap ranges (last origination – first extinction) for suites of long-ranging fossil species so as to provide maximum constraint upon the ages of the strata that contain them.

6. The constancy of a half-life for any given unstable isotope implies that radiometric decay is nonlinear (initially more rapid, then slowing down through time).

7. Earth is a geologically active planet; Earth materials are continually recycled, so that the age of the Earth is likely significantly older than the oldest known rocks (~ 4.0 Ga). The widely cited 4.6 Ga age of the Earth arises from radiometric dating of meteorites (orbital characteristics of the solar system imply contemporaneous formation of the Sun, planets, and asteroids).

Answers to review questions

1. How was the duration of a second originally defined? How is it defined now?

Originally, a second was 1/86,400th of the Earth's rotational period, as approximated by the position the Sun (i.e., 1/86,400th the time elapsed between "high noon" on successive days). The modern definition of a second is the time it takes the magnetic polarity of a cesium atom to experience 9,192,631,770 reversals.

2. How do sidereal and tropical years differ?

A sidereal year is the time it takes Earth to traverse through its orbit until it reaches the same position it originally had with respect to the "fixed" background of stars (i.e., one full revolution). Human calendars attempt to chart the course of the (slightly shorter, by less than 30 minutes) tropical year, the length of time between successive vernal equinoxes. This disparity arises from the precession of the Earth's rotational axis (sometimes termed the precession of the equinoxes). Earth's geographic north pole is not only tilted away (currently 23.5°) from an ideal normal to the ecliptic (orbital) plane, but it also rotates about this normal axis with a periodicity of approximately 26,000 years. As a result, the northern spring equinox returns after slightly less than a full revolution about the Sun.

3. Why do calendars get out of sync with the seasons?

Calendars get out of sync with the seasons because the tropical year is not a whole number of days but rather 365.24219 . . . .
The problem was first addressed through the addition of a leap day (February 29) every fourth year, to make an approximation of 365.25 days. But this still proved to be too imprecise, and the seasons kept slipping. Pope Gregory (for whom the Gregorian calendar is named) instituted further refinement, eliminating years that were multiples of 100 from being leap years unless they were also multiples of 400. (Thus 2000 C.E. was a leap year, but 1900 C.E. was not). This made the year average out to 365.2425 days, which was deemed close enough for agricultural work (timing of planting and harvesting). In the twentieth century, technology allowed greater precision; ultimately the aforementioned cesium atomic clock was employed in the quest for ever more accurate timekeeping.

4. What was Steno's insight into the nature of fossils?

Unlike most of his contemporaries, Steno understood that fossils were the remains of ancient life.

5. Compare numerical age and relative age.

Relative age is a statement of ordinal timing; e.g., Ordovician sedimentary rocks were deposited after those of the Cambrian, but before those of the Silurian. A numerical age for a rock consists of a number estimating the time of the rock's formation in years before the present.

6. Describe seven principles that allow us to determine the relative ages of geologic events.

Along with fossil succession (see next question), seven physical principles are:

superposition — In an undisturbed sequence of sedimentary rocks, younger layers are deposited on top of older layers.

original horizontality — Sediments are deposited in horizontal layers; variance from this implies postdepositional deformation.

original (lateral) continuity — Spatially isolated bodies of sedimentary rock of identical lithology were once physically connected. Isolation resulted from erosion of a previously continuous planar bed.

cross-cutting relations — Features can only disrupt other features which are older than themselves: A. Igneous intrusions are younger than rocks they intrude. B. Faults are younger than the rocks and surfaces they offset. C. Erosional surfaces are younger than the rocks into which they cut.

inclusions — If a rock contains pieces of another body of rock, then it must be younger than the donor rock.

baked contacts — If a rock is in places metamorphosed as a result of contact with a hot body of magma or lava (which later crystallized to form igneous rock), the metamorphosed rock must have existed (as unaltered rock) prior to the metamorphic event.

uniformitarianism — "The present is the key to the past." Ancient geological events are expected to have been similar to those that have been historically recorded, as there is no reason to assume that the laws of nature have changed over time. For example, great bodies of mudstone preserved in the rock record are the result of long-term, slow (on a human scale) deposition of muddy sediment at the surface. Muddy (or sandy, or pebbly) strata do not intrude from below the surface, and it is clear that this has always been the case given our present knowledge of the Earth's interior. Uniformitarian views of deposition and gravity lead directly to Steno's principles (the first three above).

7. How does the principle of fossil succession allow us to better determine the relative ages of geologic events?

Species originate within geologically brief time frames, few species persist for more than a few million years, and extinction is permanent. Fossils that are, for example, uniquely found within Maastrichtian rocks can be used to illustrate temporal equivalence of isolated strata that bear these same fossil assemblages. Thus, physically isolated sedimentary sequences can be temporally correlated in the absence of datable igneous intrusions. Additionally, strata containing characteristically Danian assemblages must be younger than those with Maastrichtian fossils.

8. How does an unconformity develop? What does it tell us about the pace of geologic events?

An unconformity occurs when sedimentary layers are deposited on top of a surface of erosion or nondeposition. Within a sequence of rocks, unconformities may represent immense spans of time that are not represented by rock. Thus, the time range encompassed by a sequence of rocks containing unconformities is generally far greater than might be estimated dividing sedimentary thickness by a plausible depositional rate.

9. Describe the differences between the three kinds of unconformities.

Nonconformities arise when sedimentary strata are deposited on top of crystalline (metamorphic or igneous) rock. Disconformities occur when strata are deposited on top of an erosional surface that was horizontal at the time, so that layers above and below the unconformity are parallel. Angular unconformities arise when the new strata are deposited on top of older layers that have been tilted out of horizontality, so that, regardless of future tilting, layers on opposite sides of the unconformity are not parallel.

10. Describe two different methods of correlating rock units.

Lithologic (a.k.a. lithostratigraphic) correlation is the use of physical and chemical characteristics of rocks to determine that spatially isolated strata were once continuous (through original lateral continuity). Fossil (a.k.a. chronostratigraphic) correlation uses common index fossils to determine the approximate temporal equivalence of two bodies of rock (which may be of disparate lithologies).

11. What does the process of radioactive decay entail?

Radioactive decay is the nuclear breakdown of an unstable isotope through either: A. ejection of an alpha particle (two protons and two neutrons; this is termed "alpha decay"), B. conversion of a neutron into a proton plus an electron, the latter of which is expelled out of the nucleus by the weak force (electrons are termed beta particles, so this process is called "beta decay"), C. electron capture by a proton to form a neutron, or D. through fission of a large isotope into two smaller ones (alpha decay being a specific example of this).

12. How do geologists obtain a radiometric date? What are some of the pitfalls in obtaining a reliable one?

Geologists obtain a radiometric date by observing the ratio of unstable parent isotope to the stable daughter product of the reaction within a mineral extracted from the rock of interest and by comparing this ratio to the known half-life for the nuclear reaction. To be accurate, the following conditions must be met: 1) The rock must be young enough so that detectable amounts of parent isotope remain in the mineral, yet old enough so that sufficient amounts of daughter isotope have been produced. 2) Since the time of mineral formation, the rock must not have been reheated above the blocking temperature of the mineral under study. 3) At the time of mineral formation (more technically, the time of cooling through the blocking temperature), the concentration of daughter product within the rock was zero. 4) Since the time of mineral formation, neither parent nor daughter isotope have been introduced to the mineral crystal under study, daughter isotope has not escaped from the crystal, and the concentration of the parent isotope has not been reduced through any process other than the nuclear reaction (or reactions) under consideration.

Escape of the daughter product can sometimes be ameliorated by comparing abundance of the unstable parent isotope to that of a stable isotope of the same element, given a known ratio of stable versus unstable isotopes for the parent element prior to the inception of nuclear decay (as in radiocarbon dating).

13. Why can't we date sedimentary rocks directly?

Radiometric dates provide ages of minerals (grains or crystals) within rock. Igneous rocks form as their component minerals form and are thus datable. Sedimentary rock grains, such as sand, mud, and gravel, derive from the physical and chemical weathering of preexisting rocks at relatively low temperatures. The mineral components within sedimentary rocks are generally much older than the rocks themselves.

14. Why is carbon-14 dating used for very young rocks and archeology, but useless for dating dinosaur fossils?

　　　The half-life of carbon-14 is less than 6000 years. Excluding their likely descendants, the birds, the last dinosaurs lived 65 million years ago. No detectable amounts of carbon-14 could be found in material so ancient. (Bones and teeth are primarily mineralized with calcium phosphate but have minor amounts of calcium carbonate; dinosaur egg shells have a greater percentage of calcium carbonate, similar to the shells of living birds.)

15. How are growth rings and ice cores useful in determining the ages of geologic events?

　　　Growth rings in tree bark mark annual growth cycles and vary in thickness depending upon water availability. Over small geographic regions, patterns of wet and dry years can be matched up from tree to tree. Younger fossil trees can even be compared to very old living trees to provide a numerical age.

　　　Isotopic ratios of oxygen atoms within ice vary with Earth's global climate at the time of ice formation, so isotopic profiles from disparate localities can be compared to provide correlation as well as climatic information.

16. How are the reversals of the Earth's magnetic field useful in dating geologic events?

　　　Crystals of magnetic minerals, such as magnetite, form through extrusive volcanism and become magnetized in alignment with Earth's magnetic field at the time they cool past their Curie point. Barring subsequent reheating to the Curie point, they retain this remnant magnetization. Polarity reversals are worldwide events, and the durations of "normal" and "reversed" periods have been highly variable. Under reasonably constant rates of volcanism, the alternating thicknesses of "normal" and "reversed" layers (adduced from magnetic anomaly data) within a sequence can be compared on a global basis.

17. Why is it inappropriate to use the term "absolute age" for numerical dating?

　　　Numerical ages are measured in years before the present. This is convenient, but the present does not have any logical priority as time zero. Even numerical ages must be measured relative to some (arbitrary) event.

18. Why did early scientists think the Earth was less the 100 million years old?

　　　They accepted estimates derived from unreliable assumptions, such as Kelvin's assumption that the Earth was merely cooling and did not obtain heat from any internal sources.

19. How did the discovery of radioactivity invalidate Kelvin's assumptions about the Earth's age and also provide a method for obtaining its true age?

　　　Radioactivity is an important source of heat for the Earth, keeping it from cooling precipitously as was envisioned by Kelvin. It also allows radiometric dating, which has since been used to date Earth rocks more accurately.

20. What is the age of the oldest rocks on Earth? What is the age of the oldest rocks known? Why is there a difference?

The oldest rocks on Earth are about 4 billion years old. Meteorites are as ancient as 4.6 billion years old; this is likely the age of the entire solar system (including Earth) given the orbital characteristics of the planets. No Earth rocks are likely to be found yielding 4.6 Ga dates because the Earth was initially too hot (with heat derived from initial accretion, collisions with other solar system bodies, and differentiation to form the core, mantle and crust, as well as radioactivity) to sustain solid rock at the surface for any great length of time. Even today, Earth is geologically active in comparison to other bodies in the solar system. Erosion, subduction, and melting are processes that actively destroy rock, so the oldest rock preserved is unlikely to be the oldest rock formed.

Test bank

Questions 1–13 relate to the geologic cross-section on the top of the next page. Note that the legend lists rock types in alphabetical, rather than stratigraphic, order.

1. The surfaces named contacts #2 and #3 are examples of _____.
    A. conformable sedimentary contacts   B. faults
    C. baked contacts   D. unconformities

2. Contact #1 is an example of a(n) _____.
    A. conformable sedimentary contact   B. fault
    C. baked contact   D. unconformity

3. The oldest geologic unit in the cross-section is the _____.
    A. granite   B. marble   C. limestone   D. sandstone

4. The youngest geologic unit visible in the cross-section is the _____.
    A. granite   B. limestone   C. conglomerate   D. marble

5. The marble rind surrounding the granite pluton must be younger than the limestone according to the principle of _____.
    A. superposition   B. baked contacts
    C. original horizontality   D. original continuity

6. Concerning relative ages of the shale and sandstone, _____.
    A. the shale must be older, according to the principle of superposition
    B. the sandstone must be older, according to the principle of superposition
    C. the shale must be older, according to the principle of components
    D. their relative ages cannot be determined from the information given

7. Concerning relative ages of the granite and sandstone, _____.
    A. the granite must be older, according to the principle of superposition
    B. the sandstone must be older, according to the principle of superposition
    C. the granite must be older, according to the principle of components
    D. their relative ages cannot be determined from the information given

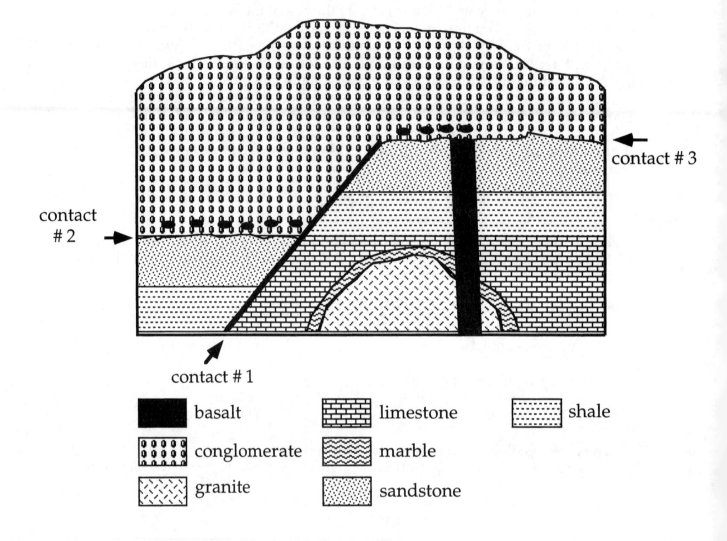

8. Basaltic clasts within the basal layers of the conglomerate imply which of the following statements (choose the most correct answer)?
    A. The conglomerate must be the older of the two.
    B. The basalt must be the older of the two.
    C. The section must not have been overturned by tectonic activity at any time.
    D. B and C are both correct.

9. Basaltic clasts within the conglomerate have been radiometrically dated to 50 million years ago (Eocene epoch of the Tertiary period). Is this a reliable age for the conglomerate?
    A. Yes.
    B. No, this age is likely too old.
    C. No, this age is likely too young.
    D. No, basalt never contains minerals bearing radioactive isotopes.

10. Concerning the relative ages of the basalt and the fault, _____.
    A. the fault must be older, according to the principle of cross-cutting relationships
    B. the basalt must be older, according to the principle of cross-cutting relationships
    C. the basalt must be older, according to the principle of original horizontality
    D. their relative ages cannot be determined from the information given

11. The basalt body is best described as a _____.
    A. pluton   B. sill   C. dike   D. laccolith

12. Contact #3 is both _____ at various points.
    A. a conformable contact and an unconformity
    B. a fault and an unconformity
    C. a nonconformity and an angular unconformity
    D. a disconformity and a nonconformity

13. Which pair of contacts once formed a continuous surface?
    A. contacts #1 and #2
    B. contacts #2 and #3
    C. contacts #1 and #3
    D. This was never true for any pair of labeled contacts.

14. James Hutton, the "father of geology," put forth the principle of _____.
    A. superposition   B. original continuity
    C. original horizontality   D. uniformitarianism

15. Period names on the geologic time scale, such as Devonian and Permian, provide examples of _____.
    A. relative age   B. numerical age

16. In a undisturbed sequence of sedimentary rocks, younger layers overly older layers, according to the principle of _____.
    A. superposition   B. original continuity
    C. original horizontality   D. uniformitarianism

17. If the lithology and fossil content of two bodies of rock on opposite sides of a canyon are identical, then these remaining outcrops were likely physically connected at one time and formed part of an extensive, sheet-like layer of rock. This idea summarizes the principle of _____.
    A. superposition   B. original continuity
    C. original horizontality   D. uniformitarianism

18. Which of the following geologic principles is a direct result of gravity?
    A. baked contacts   B. cross-cutting relationships
    C. original horizontality   D. inclusions

19. Which of the following geologic principles is **not** a result of gravity?
    A. original horizontality
    B. cross-cutting relationships
    C. original continuity
    D. superposition

20. **Relative** ages expressed on the geologic time scale primarily resulted from the study of _____.
    A. fossil content and spatial relationships among igneous rocks
    B. fossil content and spatial relationships among sedimentary rocks
    C. radiometric dating of igneous rocks
    D. radiometric dating of sedimentary rocks

21. **Numerical** ages for boundaries between time units on the geologic time scale primarily resulted from the study of _____, in conjunction with relative age data.
    A. fossil content and spatial relationships among igneous rocks
    B. fossil content and spatial relationships among sedimentary rocks
    C. radiometric dating of igneous rocks
    D. radiometric dating of sedimentary rocks

22. Within the world's sedimentary rocks, fossils _____.
    A. are rarely, if ever, found
    B. are randomly distributed
    C. occur in an ordered sequence

23. The surface below sedimentary rocks that overlie igneous or metamorphic rocks is termed a(n) _____.
    A. disconformity
    B. angular unconformity
    C. nonconformity

24. Buried erosional surfaces between parallel sedimentary strata are termed _____.
    A. disconformities
    B. angular unconformities
    C. nonconformities

25. Which method of correlation is more reliable for determining age equivalence among bodies of rock that are physically isolated by vast distances?
    A. lithologic correlation
    B. fossil correlation

26. Which eon of geologic time is represented by rocks containing abundant shelly fossils?
    A. Archean
    B. Phanerozoic
    C. Proterozoic

27. Two atoms of a single element that differ in number of neutrons are said to represent two distinct _____ of that element.
    A. isomers
    B. isotopes
    C. isotherms
    D. atomic species

28. What proportion of a radioactive isotope is expected to remain in an unaltered (unreacted) state after the passage of three half-lives?
    A. one-third
    B. three-halves
    C. one-eighth
    D. one-sixth

29. Precisely speaking, a measured radiometric age for a mineral crystal within an igneous rock denotes the amount of time that has passed since the _____.
    A. atoms within the crystal were part of a body of molten magma
    B. crystal solidified
    C. temperature of the crystal became equal to the Curie point for the mineral
    D. temperature of the crystal became equal to the blocking temperature for the mineral

30. Radiometric dates applied to sedimentary rocks produce ages that are _____.
    A. just as accurate as when the technique is applied to igneous rocks.
    B. uniformly too young (post-date sedimentary deposition)
    C. uniformly too old (pre-date sedimentary deposition)

31. The correct answer to question #30 illustrates the principle of _____.
    A. baked contacts
    B. cross-cutting relationships
    C. original horizontality
    D. inclusions

32. Key beds are _____.
    A. layered rocks that are most useful for lithologic correlation
    B. sandstone bodies with unique "keyhole" sedimentary structures
    C. sedimentary rocks that contain a greater quantity of fossils than adjacent layers

33. Dendrochonology involves dating of historic and geologic events through the study of _____.
    A. growth layers in shells
    B. oxygen isotope profiles in glacial ice
    C. remnant magnetism in iron-rich minerals
    D. annual growth rings in trees

34. Magnetostratigraphy takes advantage of the fact that _____.
    A. Earth's magnetic field only arose 10,000 years ago.
    B. Earth's magnetic field has been uniformly strong, with constant polarity, throughout Earth history
    C. Earth's magnetic field has experienced numerous polarity reversals, with periods of normal and reversed polarity of varying lengths

35. The age of the Earth cannot be reliably estimated from sediment thicknesses because _____.
    A. sedimentation rates are not likely to remain constant at any one locality over all of Earth's history
    B. sedimentary rocks can be metamorphosed or melted
    C. locally, much of Earth history is represented by unconformities between strata rather than the strata themselves
    D. All of the above are possible reasons.

36. Attempts to measure the age of the Earth by extrapolating modern riverine influx rates into the past to produce modern marine salinity from an initially freshwater ocean failed to take into account that _____.
    A. salt is removed through precipitation, balancing riverine input
    B. dissolved salts from weathering of rock only rarely are transported all the way to the ocean
    C. salts derived from weathering of rock are insoluble in water

37. Why is radiocarbon dating only rarely applied in geological work?
    A. No substances on Earth contain significant amounts of carbon-14.
    B. The half-life of carbon-14 is so long that it is effectively a stable isotope.
    C. The half-life of carbon-14 is so short that it can only be used to date materials that are less than 70,000 years old.

38. The principle of uniformitarianism implies that _____.
    A. all ancient landforms and geography looked identical to the modern world
    B. physical processes, such as erosion, weathering, and deposition, which we observe today, were also active in the geologic past
    C. igneous, metamorphic, and sedimentary rocks are uniformly distributed at Earth's surface

39. Uniformitarianism is succinctly summarized by which phrase?
    A. The future is the key to the present.
    B. The present is the key to the past.
    C. The past is the key to the present.
    D. The present is the key to the future.

40. Which statement best summarizes the development of the geologic time scale?
    A. Numerical ages for rocks were known well before the relative sequence of sedimentary layers was established.
    B. Relative ages for sedimentary strata were known well before accurate numerical dates for these rocks could be provided.
    C. Names of relative ages (such as Silurian) and accurate numerical dates for these ages appeared at about the same time.

Chapter 13
A Biography of Earth

Learning objectives

1. Students should know the approximate age of the Earth, how this date was obtained, and why it is substantially older than the oldest known rocks.

2. They should have a general idea about Hadean temperatures, the potential sources of heat early in Earth's history, and the thin, transient nature of the earliest crust. They should be able to distinguish the earliest, volcanically dominated atmosphere from today's.

3. They should know and understand: A. The contrast between early Archean tectonics (small protocontinents surrounded by greenstone seas) and more modern Proterozoic conditions (development of larger oceanic plates and extensive stable cratons). B. The appearance of the first life on Earth (3.8 Ga); the importance of cyanobacteria geologically (stromatolite formation, oxygenation of oceans and atmosphere). C. The formation of supercontinents Rodinia (1 Ga) and Pangaea (250 Ma); evidence for world-wide glaciation in the late Proterozoic. D. The Cambrian explosion of animal life and the development of a complex food web on the basis of predation. E. The timing of orogenic events that produced the Appalachians (Taconic, Acadian, Alleghenian) and Rockies (Nevadan, Sevier, Laramide); and the positions of the volcanic arc and fold-and-thrust belt relative to the cratonic interior. F. Major events in the history of life (from an admittedly biased vertebrate perspective): first fishes, first terrestrial vertebrates, the age of dinosaurs, and approximate timing of the origins of birds (Jurassic) and mammals (Triassic).

Summary from the text

Earth formed about 4.6 billion years ago. During the first 600 million years of its history, the Hadean eon, the planet was so hot that its surface was a magma ocean. We have no record for the radiometric clock to work, but we can gain insight into it by studying the Moon.

The Archean eon began about 4.0 Ga, when the Earth had cooled sufficiently for permanent continental crust to form. The crust assembled out of volcanic arcs and hot-spot volcanoes that were too buoyant to subduct. Water oceans and stable continental blocks called cratons also formed. The atmosphere contained little oxygen, but the first life forms, bacteria, appeared, perhaps along mid-ocean ridges.

In the Proterozoic eon, which began at 2.5 Ga, Archean cratons collided and sutured together along large orogenic belts. Island volcanic arcs and hot-spot volcanoes plastered onto the margins of continents, creating large Proterozoic cratons whose interiors were far from the orogenic belts. Photosynthesis by organisms added oxygen to the ocean to form banded-iron formations. Eukaryotic cells appeared at 1.5 Ga, greatly increasing the amount of oxygen in the atmosphere. With such an atmosphere, more complex creatures could develop, and by the end of the Proterozoic, complex but shell-less marine invertebrates populated the planet. A supercontinent called Rodinia, which reorganized to form another supercontinent called Pannotia, had accumulated most continental crust by the end of the Proterozoic eon.

At the beginning of the Paleozoic era, about 540 Ma, Pannotia broke apart, yielding several smaller continents. The sea level rose and fell a number of times, creating sequences of strata in continental interiors. Continents began to collide and coalesce again, leading to orogenies (including the Taconic, Acadian, and Alleghenian of the Appalachian Mountains) and, by the end of the era, another supercontinent, Pangaea. Invertebrate organisms evolved shells, and life underwent diversification. Milestones in evolution include the appearance of trilobites, brachiopods, and large predators in the Cambrian period: jawless fish in the Ordovician period; land plants in the Silurian period; the first land animals (insects and amphibians), jawed fish, and seed plants in the Devonian period; and reptiles and gymnosperm trees in the Carboniferous period.

In the Mesozoic era, Pangaea broke apart and the Atlantic Ocean formed. Convergent-boundary tectonics dominated along the western margin of North America, causing orogenic events like the Sevier and Laramide orogenies. Dinosaurs appeared in late Triassic time and became the dominant land animals through the Mesozoic era. During the Cretaceous period, the sea level was very high, and the continents flooded. Angiosperms (flowering plants) appeared at this time, along with modern fish. A huge mass extinction event (the K-T boundary event), which wiped out the dinosaurs, occurred at the end of the Cretaceous period, probably because of the impact of a large bolide with the Earth.

In the Cenozoic era, continental fragments of Pangaea began to collide again. The collision of Africa with Europe and of India with Asia formed the Alpine-Himalayan orogen. Convergent tectonics has persisted along the margin of South America, creating the Andes, but ceased in North America when the San Andreas Fault formed. Rifting in the western United States during the Cenozoic era produced the Basin and Range Province. Various kinds of mammals took the place of dinosaurs on Earth, and the human genus, *Homo*, appeared and evolved throughout the radically shifting climate of the Pleistocene epoch.

The end of the Earth will probably happen 5 billion years in the future, when the Sun runs out of nuclear fuel, transforms briefly into a red giant, and vaporizes our planet.

Answers to review questions

1. List some methods by which geologists study the past.

Examining rocks and rock sequences, their fossil continent, and structures (which may reveal information concerning conditions at the time of rock formation or else subsequent deformation). Sediment, ice cores, and tree rings additionally provide information for the most recent history.

2. Why are there no rocks on Earth that yield radiometric dates older than 4 billion years?

See review question #20 from Chapter 12.

3. Describe the condition of the crust, atmosphere, and oceans during the Hadean eon.
   The crust would have been largely ultramafic magma and rock (which may have been subject to remelting). The atmosphere lacked free oxygen and would have been dominated by volcanically emitted gases (water vapor, nitrogen, carbon dioxide). For at least part of the Hadean, temperatures were likely too hot to sustain a liquid ocean.

4. Describe five principal rock types found in Archean protocontinents. Under what kind of environmental conditions did each type form?
   Gneiss — collision-induced metamorphism of protocontinental rock.
   Greenstone — metamorphosed oceanic basalt.
   Granite — crustal melting and subsurface cooling.
   Graywacke — "dirty sandstones"; terrigenous mud and clay carried to and deposited in deep ocean basin.
   Chert — chemically precipitated from ocean water.

5. What are stromatolites? How do they form?
   Stromatolites are sedimentary mounds created by the presence of sticky cyanobacterial mats which trap sediment. They are finely laminated as a result of the cyanobacteria working their way upward to recolonize the surface after they become too deeply buried in trapped sediment to effectively photosynthesize.

6. How did the atmosphere and tectonic conditions change during the Proterozoic eon?
   The atmosphere became oxygenated (and depleted of some carbon dioxide) as a result of photosynthesis. Plate tectonic conditions became more similar to what is observed today, with larger oceanic plates and the development of broad, stable cratons in continental interiors.

7. How do banded-iron formations (BIFs) tell us about atmospheric oxygen levels?
   BIFs formed when dissolved oxygen in the seas oxidized soluble (ferrous) iron to produce insoluble (ferric) iron, which precipitated as iron oxide. The cessation of BIFs later in the Proterozoic implies that dissolved oxygen levels would have built up until release through effervescence to the atmosphere would have been inevitable. BIF cessation thus provides timing on the inception of atmospheric oxygenation.

8. What evidence do we have that the Earth nearly froze over twice during the Proterozoic eon?
   Glacial deposits have been found in sedimentary rocks representing sea-level environments along the paleoequator.

9. How did the Cambrian explosion of life change the nature of the living world?
   These first abundant animals complicated the food web through the introduction of predation and brought increased bioturbation of sediment as

well as the first biogenic reefs. Carnivores likely induced selection pressure favoring numerous defensive structures (thick shells, spines) and behaviors (burrowing, active swimming).

10. How did the Taconic and Acadian orogenies affect the east coast of North America?
    They deformed local strata and uplifted eastern mountains in the region of the modern Appalachians; these mountains eroded to form thick sedimentary deposits.

11. How did the Alleghenian and Ancestral Rocky orogenies affect North America?
    The Alleghenian orogeny provided the final uplift to produce the Appalachian Mountains. This powerful collision between Laurentia and Gondwana influenced fault movement in what is now western North America. Uplifted western blocks (the Ancestral Rockies) eroded and subsequently lithified to form the red sandstone typical of the region.

12. Compare the typical sedimentary deposits of the early Paleozoic greenhouse Earth with those of the late Paleozoic/early Mesozoic icehouse Earth.
    In North America, early and mid-Paleozoic greenhouse conditions led to widespread epicontinental seas, producing sedimentary sequences dominated by fossiliferous limestone. Late Paleozoic cooling brought a retreat of the seas and led to nearshore, deltaic, and freshwater deposits of sand, mud, and organic carbon (from accumulation of swamp vegetation) to form sandstone, mudstone, shale, and coal.

13. Describe the plate tectonic conditions that led to the formation of the Sierran arc and the Sevier fold-thrust belt.
    North America had a convergent plate boundary to the west which induced collisions with microcontinents as well as subduction of oceanic crust, which partially melted to form the Sierran volcanic arc. Compression induced the folds and thrust faults just interior to the arc itself.

14. How did the plate tectonic conditions of the Laramide orogeny differ from more typical subduction zones?
    Faulting associated with the Laramide event uplifted deep basement rock to produce the Rocky Mountains, whereas earlier orogenies produced only the more superficial faults of the fold-and-thrust belts behind the arc. The volcanic arc also became discontinuous, with volcanism ceasing in the central Rocky Mountains region (as the result of a shallow dip of the subducting slab).

15. What caused the flooding of the continents during the Cretaceous period?
    Rapid Cretaceous rates of submarine volcanism thickened the Mid-Ocean Ridge and produced numerous hot spots, which displaced water onto the continents. Additionally, standing ice could not accumulate anywhere on Earth in the associated greenhouse warmth.

16. What could have caused the K-T extinctions?
    The extraterrestrial impact that produced the Chicxulub crater was certainly a factor. Large-scale volcanic activity (evidenced by the basalts of the Deccan traps of India) also characterized the latest Cretaceous and may have induced climatic change or instability.

17. What new continents formed as a result of the breakup of Pangaea?
    North America, South America, and Africa are examples (answers may vary).

18. What are the causes of the uplift of the Himalayas and the Alps?
    The Alps arose from Europe's collision with Africa; the Himalayas arose from India's collision with Asia.

19. What events led to the end of the Mesozoic greenhouse and the development of glaciers on the Arctic and Antarctic during the Cenozoic era?
    Reduced submarine volcanic rates, isolation of Antarctica over the south pole, restricted circulation of Arctic waters by Eurasia and North America, and the spread of grasses led to cooling during the Cenozoic. Grasses utilize carbon dioxide during photosynthesis (as do all other plants), and additionally they are difficult for herbivores to eat (tough siliceous stems). Grasslands additionally increase the albedo of the Earth (they're lighter in color than forests, so they lead to greater reflection of sunlight), and they thrive in relatively cool, dry settings.

20. Describe some possible scenarios for the future evolution of the Earth.
    Unless global warming from increased greenhouse gas concentrations is sufficient to melt polar ice deposits, ice sheets should return to North America in a few thousands of years. Ultimately, however, Earth will be destroyed by catastrophic impact or the onset of the red giant stage of our Sun's development.

Test bank

1. The Earth is approximately how old?
    A. 15 billion years
    B. 4.6 billion years
    C. 40 million years
    D. 6,000 years

2. The Hadean is a time in Earth history when _____.
    A. the first abundant shelly organisms appear in the fossil record
    B. Earth's interior was so hot that a solid outer crust, if present, was likely being extensively re-melted
    C. stable continental interiors, termed cratons, first formed
    D. the dinosaurs first appeared and came to dominate large-scale terrestrial life

3. The Cambrian period is a time in Earth history when _____.
    A. the first abundant shelly organisms appear in the fossil record
    B. Earth's interior was so hot that a solid outer crust, if present, was likely being extensively re-melted

C. stable continental interiors, termed cratons, first formed
D. the dinosaurs first appeared and came to dominate large-scale terrestrial life

4. The Mesozoic era is a time in Earth history when _____.
   A. the first abundant shelly organisms appear in the fossil record
   B. Earth's interior was so hot that a solid outer crust, if present, was likely being extensively re-melted
   C. stable continental interiors, termed cratons, first formed
   D. the dinosaurs first appeared and came to dominate large-scale terrestrial life

5. Towards the end of the Archean eon, _____.
   A. the first abundant shelly organisms appear in the fossil record
   B. Earth's interior was so hot that a solid outer crust, if present, was likely being extensively re-melted
   C. stable continental interiors, termed cratons, first formed
   D. the dinosaurs first appeared and came to dominate large-scale terrestrial life

6. Greenstones are metamorphosed sedimentary rocks; as sediments, they were deposited in the oceans during the _____.
   A. Hadean   B. Paleozoic
   C. Proterozoic   D. Archean

7. Which gas found in today's atmosphere was absent in the Hadean's and Archean's?
   A. nitrogen   B. water vapor
   C. oxygen   D. carbon dioxide

8. Greenhouse climate on Earth has been associated with relatively high atmospheric levels of which gas?
   A. nitrogen   B. water vapor
   C. oxygen   D. carbon dioxide

9. Rodinia and Pangaea are supercontinents that assembled during the _____, respectively.
   A. Hadean and Archean   B. Archean and Proterozoic
   C. Proterozoic and Paleozoic D. Paleozoic and Mesozoic

10. The Taconic, Acadian, and Alleghenian orogenic events all led to uplift in the region of the modern _____.
    A. Rockies   B. Appalachians
    C. Alps   D. Himalayas

11. The volcanic arc along the western coast of North America during the Cretaceous period resulted from subduction of _____.
    A. continental lithosphere underneath continental lithosphere
    B. continental lithosphere underneath oceanic lithosphere
    C. oceanic lithosphere underneath oceanic lithosphere
    D. oceanic lithosphere underneath continental lithosphere

12. Where today would you expect to find rocks deposited by late Paleozoic glaciers?
    A. United States   B. western Russia
    C. central Europe   D. India

13. The Himalayas were produced by the collision of _____ with southern Asia during the Cenozoic.
    A. India   B. Australia
    C. Africa   D. South America

14. Compared to the Silurian and Devonian periods, Earth during the Permian was generally _____.
    A. warmer   B. cooler

15. In the interior of North America, sedimentary sequences dominated by fossiliferous limestone were deposited during the _____.
    A. early and middle Paleozoic   B. late Paleozoic
    C. early and middle Cenozoic   C. late Cenozoic

16. The small cells of bacteria lack nuclei and are termed _____.
    A. prokinetic   B. prokaryotic   C. eukaryotic   D. analgesic

17. Most DNA in the cells of animals and plants is housed within a nucleus; their cells are termed _____.
    A. prokinetic   B. prokaryotic   C. eukaryotic   D. analgesic

18. Of the following choices, rates of spreading along the Mid-Ocean Ridge were fastest during the _____.
    A. Oligocene   B. Eocene
    C. Permian   D. Cretaceous

19. *Anomalocaris* was a marine carnivore that lived during the _____.
    A. Cambrian   B. Devonian
    C. Permian   D. Cretaceous

20. The Cambrian explosion not only produced an abundance of new animal species, but also gave rise to the first large shells, spines, and active swimmers. This suggests the important influence of _____ in driving animal evolution.
    A. photosynthetic organisms   B. atmospheric oxygen
    C. carnivorous predators   D. filter-feeding organisms

21. The oldest terrestrial insect fossils are found in rocks from the _____ period.
    A. Cambrian   B. Devonian   C. Permian   D. Triassic

22. The first terrestrial vertebrates (with an amphibian life cycle) are represented by fossils that are late _____ in age.
    A. Cambrian   B. Devonian   C. Permian   D. Triassic

23. Dinosaurs first appeared during the _____ period.
    A. Cambrian   B. Devonian   C. Permian   D. Triassic

24. The first mammals appeared during the _____ period.
    A. Cambrian   B. Devonian   C. Permian   D. Triassic

25. The eon represented by abundant deposits of strata with shelly fossils is the _____.
    A. Archean   B. Hadean
    C. Proterozoic   D. Phanerozoic

26. A shield is _____.
    A. a low-lying region where Precambrian rocks crop out
    B. a low-lying region where Precambrian rocks are covered by Phanerozoic sedimentary strata
    C. a region of compressed rock located behind a volcanic arc at a convergent continental margin
    D. synonymous with "craton"

27. A platform is _____.
    A. a low-lying region where Precambrian rocks crop out
    B. a low-lying region where Precambrian rocks are covered by Phanerozoic sedimentary strata
    C. a region of compressed rock located behind a volcanic arc at a convergent continental margin
    D. synonymous with "craton"

28. Geographically, a shield and surrounding platform together are _____.
    A. a low-lying region where Precambrian rocks crop out
    B. a low-lying region where Precambrian rocks are covered by Phanerozoic sedimentary strata
    C. a region of compressed rock located behind a volcanic arc at a convergent continental margin
    D. synonymous with "craton"

29. A fold-and-thrust belt is _____.
    A. a low-lying region where Precambrian rocks crop out
    B. a low-lying region where Precambrian rocks are covered by Phanerozoic sedimentary strata
    C. a region of compressed rock located behind a volcanic arc at a convergent continental margin
    D. synonymous with "craton"

30. The late Proterozoic Grenville orogen is an event related to _____.
    A. uplift of the modern Rocky Mountains
    B. uplift of the ancestral Rocky Mountains
    C. assembly of the supercontinent Pangaea
    D. assembly of the supercontinent Rodinia

31. The late Paleozoic Hercynian orogen is an event related to _____.
    A. uplift of the modern Rocky Mountains
    B. uplift of the ancestral Rocky Mountains
    C. assembly of the supercontinent Pangaea
    D. assembly of the supercontinent Rodinia

32. The earliest forests containing woody trees appeared during the _____.
    A. late Proterozoic   B. early Paleozoic (Ordovician)
    C. middle Paleozoic (Devonian)   D. late Paleozoic (Permian)

33. Tree-sized relatives of the living ferns, club mosses, and scouring rushes were common in environments termed coal swamps during the _____ period.
    A. Ordovician   B. Carboniferous
    C. Jurassic   D. Cretaceous

34. Permian, Triassic, and Jurassic forests were dominated by large _____.
    A. angiosperms (flowering plants)
    B. mosses
    C. conifers and cycads
    D. club mosses and scouring rushes

35. Which flying reptiles, first appearing during the Triassic, were the earliest flying vertebrates?
    A. birds   B. aerogekkos
    C. pterosaurs   D. phytosaurs

36. The Western Interior Seaway reached its greatest level of continental inundation during the _____ period.
    A. Ordovician   B. Carboniferous
    C. Jurassic   D. Cretaceous

37. *Tyrannosaurus rex*, one of the largest carnivorous dinosaur species, lived during the _____ period.
    A. Ordovician   B. Carboniferous
    C. Jurassic   D. Cretaceous

38. The Cretaceous-Tertiary (K-T) boundary event took place how many millions of years ago?
    A. 540   B. 245   C. 65   D. 25

39. An iridium-rich layer of clay, glass spherules, and shocked quartz grains strongly suggest that the K-T boundary extinctions were at least in part caused by _____.
   A. a long trend of global warming
   B. a long trend of global cooling
   C. worldwide explosive volcanism
   D. the impact of a bolide (meteoroid or comet)

40. The Chicxulub crater is located in which continent?
   A. North America   B. South America
   C. Asia   D. Africa

41. Convergence along the western margin of North America during the Cenozoic gave way to rifting forces which created the characteristic topography of the _____.
   A. Coastal Ranges   B. Basin and Range Province
   C. Colorado Plateau   D. Columbia Plateau

42. The Miocene epoch saw the spread of what new vegetative zone?
   A. grasslands   B. angiosperm forests
   C. gymnosperm forests   D. tree fern forests

43. North and South America have been physically connected for the last _____.
   A. 65 million years   B. 5 million years
   C. 2.5 million years   D. 75,000 years

44. The most recent interval of Quaternary glaciation gave way to warmer, interglacial conditions _____ ago.
   A. 1 million years   B. 200,000 years
   C. 75,000 years   D. 11,000 years

45. During the Cenozoic era, most large-bodied terrestrial animals inhabiting regions of temperate climate have been _____.
   A. birds   B. dinosaurs   C. mammals   D. crocodiles

46. When did the human genus, *Homo*, first appear on the Earth?
   A. 200 million years ago B. 2.4 million years ago
   C. 100,000 years ago   D. 10,000 years ago

47. What closely related genus has a somewhat more ancient fossil record than *Homo*?
   A. *Australopithecus*   B. *Gorilla*
   C. *Tyrannosaurus*   D. *Thylacosmilus*

48. Humans belong to a lineage of mammals called the _____.
   A. carnivores   B. primates
   C. odd-toed ungulates   D. edentates

49. The Sun is expected to transform into a red giant star in approximately _____ years.
    A. 1 million   B. 1 billion   C. 5 billion   D. 100 billion

50. In its final stage, the Sun is expected to become a _____ .
    A. normal sequence star   B. red giant
    C. white dwarf   D. neutron star

Chapter 14
Squeezing Power from a Stone: Energy Resources

Learning objectives

1. Students should know the major energy resources found within rocks and sediments (oil, natural gas, coal, and uranium), how they form, how they are recovered, and how they are utilized for our energy needs.

2. Oil and gas are commonly produced together in the heat of burial as a result of the alteration of organic-rich black shale. The organic matter within the shale is derived from the remains of plankton, which settled to the bottom of the water in an oxygen-poor environment (retarding bacterial decay rates). Gas (a short-chain hydrocarbon) is less dense and more volatile than oil (which is composed of longer carbon chains) and forms at somewhat higher temperatures (and thus is found in the absence of oil at greater depths and temperatures).

3. A recoverable quantity of oil requires that an organic-rich source rock be subjected to temperatures within the oil window. Additionally at least two other bodies of rock must be present. Directly above the source rock, a highly permeable reservoir rock must reside. Oil from the source will migrate upward through the pores of the reservoir rock because it is less dense than the water that is also found within the pores. An impermeable (seal) rock must lie above the reservoir rock, so as to keep the oil from escaping to the surface. Further, economically feasible bodies of oil require a geometric configuration that allows oil to pool up in a restricted space, bounded by seal rock so that it cannot reach the surface. Together, this configuration and the seal rock form what is termed an oil trap.

4. Coal is the altered remains of ancient plants that lived in swampy environments. The acidic water of swamps retards decay, allowing wood and leafy matter to accumulate and compress to form peat. Later, deep burial of peat at great temperatures burns off volatile materials to leave behind greater amounts of organic carbon. Grades of coal are ranked, on the basis of their carbon content, from lignite (low rank) through bituminous to anthracite (high rank), which has the greatest proportion of carbon and produces the greatest amount of heat per unit volume of material burned. Problems associated with coal mining include the incidence of acid mine drainage and acid rain because most coal contains sulfur and is found in association with pyrite (iron sulfide).

5. Most mined uranium has been hydrothermally concentrated in veins within granite plutons. The isotope used in fission reactors is relatively sparse within naturally occurring ore (uranium oxide) and must be enriched (concentrated). Fission reactors provide abundant energy, but the storage of radioactive waste is a major environmental concern.

6. The world's supply of oil, a nonrenewable fossil fuel, is finite and at current rates of usage will last for no more than 150 years. Alternative sources of energy (geothermal, wind, solar, increased usage of coal) will have to be explored further in the not too distant future.

Summary from the text

People's energy needs, which have increased dramatically during the past two centuries, are being met by resources from geological materials.

Usable energy ultimately comes from five sources: the Sun, gravity, nuclear fission, the heat stored in the Earth, and chemical reactions.

Energy resources come in a variety of forms: energy directly from the Sun; energy from tides, flowing water, or wind; energy stored by photosynthesis (either in contemporary plants or in fossil fuels); energy from inorganic chemical reactions; energy from nuclear fission; and geothermal energy (from Earth's internal heat).

Oil, gas, and coal are fossil fuels: they store solar energy that came to Earth long ago.

Oil and gas are hydrocarbons, a type of organic chemical. The viscosity and volatility of a hydrocarbon depend on the length of its molecules. When hydrocarbons burn, they react with oxygen to release carbon dioxide, water, and heat.

Oil and gas originally develop from the bodies of plankton, which settle out in a quiet-water, oxygen-poor depositional environment and form black organic shale. Later, chemical reactions at elevated temperatures convert the dead plankton into kerogen, then oil. Shale containing oil is oil shale.

In order to create a usable oil reserve, oil must migrate from a source rock (oil shale) into a porous and permeable rock called a reservoir rock. Oil rises because it is more buoyant than water. Unless the reservoir rock is overlain by an impermeable seal rock, the oil will escape to the ground surface. The subsurface configuration of strata that leads to the entrapment of oil in a good reservoir rock is called an oil trap.

Companies invest huge amounts of money to explore for oil. Geologists use a combination of surface maps, cross-section diagrams, and seismic-reflection profiling to create an image of the subsurface. Such images help them to identify traps.

Coal is formed from plant debris deposited in ancient swamps or forests. Coal-forming environments were particularly common during the Carboniferous period.

For coal to form, the plant material must be deposited in an oxygen-poor environment, so that it does not completely decompose. Compaction near the ground surface creates peat, which, when buried deeply and heated, transforms into coal, which has a high concentration of carbon.

Coal is classified into ranks, on the basis of the amount of carbon it contains, as either lignite (low rank), bituminous (higher rank), or anthracite (still higher rank). If temperatures are too high, coal completely decomposes, and the carbon recrystallizes to form graphite.

Coal occurs in beds, interlayered with other sedimentary rocks. Coal beds can be mined by either strip mining or underground mining.

Nuclear power plants generate energy by using the heat released by the nuclear fission of radioactive elements. The heat turns water into steam, and the steam drives turbines.

Uranium atoms are too big to fit into the minerals making up the mantle, so the atoms tend to concentrate in magmas that rise and ultimately form granite plutons in the crust.

Some economic uranium deposits occur as veins in igneous rock bodies, some occur in sedimentary beds composed of grains eroded from the igneous rocks, and some occur in minerals precipitated from groundwater that passed through uranium-bearing rocks.

Nuclear reactors must be carefully controlled to avoid overheating or meltdown. The disposal of radioactive nuclear waste can create environmental problems.

Geothermal energy uses Earth's internal heat to transform groundwater into steam that drives turbines; hydroelectric power uses the potential energy of water, and solar energy uses solar cells to convert sunlight to electricity.

We now live in the Oil Age, but oil supplies are expected to last for only a century or so. Tar sand, oil shale, gas, and coal may be our next main sources of energy.

Most energy resources have environmental consequences. Oil spills pollute the landscape, the sulfur associated with some fuel deposits causes acid mine runoff and acid rain, and the burning of hydrocarbons produces smog and may cause global warming.

Answers to review questions

1. Before the eighteenth century, where did people get most of their energy and fuel? How did this change during the Industrial Revolution?

Energy was harnessed from animal power, wind, and running water; wood and dried dung were used as fuel for fires. During the Industrial Revolution, wood in Europe was in short supply, and coal came to be used extensively for the first time.

2. What are the five fundamental sources of energy? How do they become eight sources of usable energy?

The five fundamental sources are nuclear fusion (within the Sun), nuclear fission (within radioactive isotopes), gravitational potential energy, chemical potential energy, and geothermal energy (the Earth's internal heat). Eight commonly utilized sources derived from these are 1) direct harnessing of solar radiation; 2) harnessing of tides (driven by the Moon's gravitational pull); 3) harnessing of the wind (driven by solar heat and gravity); 4) consuming photosynthetic plants (through digestion or combustion; plants attain their energy through photosynthesis aided by solar radiation); 5) deriving energy from chemical reactions; 6) burning fossil fuels (remains of the biosphere which ultimately attained energy from photosynthesis); 7) harnessing energy released through fission of unstable isotopes; and 8) harnessing the internal heat of the Earth.

3. How does the length of a hydrocarbon chain affect its viscosity and volatility?

Short hydrocarbon chains are less viscous and more volatile; longer hydrocarbon chains are more viscous and less volatile.

4. What is the source of the organic material in oil?
Ancient plankton.

5. What is the oil window, and why does oil form only there?
The oil window is the narrow range of temperatures in which oil forms from organic matter.

6. How is organic matter trapped and transformed to create an oil reserve?
The organic matter derived from the remains of plankton are incorporated into sediment at the bottom of the ocean to form organic-rich mud. After lithification, it becomes bound into a rock termed black shale (largely composed of organic matter and clay minerals). In the burial environment, if the remains are subjected to temperatures that are within the oil window, oil is formed. Oil is less dense than the water (which coexists with the oil in subterranean pores within rocks and sediments), so it has a tendency to float upward to the surface. In order to produce an oil field that can be commercially exploited, a porous and permeable rock (termed a reservoir) such as sandstone must reside above the source rock (oil-producing shale). The shales that produce oil do not conduct liquids rapidly, hence the need for a separate reservoir into which oil will flow. In order to keep the oil from escaping to the surface, an oil trap is necessary. The trap consists of an impermeable seal rock (such as a shale) which blocks the flow of oil, along with a geometric configuration (such as an anticline, dome, or fault) that causes the oil to migrate into a relatively confined volume.

7. Why isn't oil shale widely used as an energy source at the present time?
Obtaining oil directly from oil shale is prohibitively expensive, takes input heating energy, and produces small returns from voluminous quantities of rock.

8. How do porosity and permeability affect the oil-bearing potential of a rock?
Rocks with high porosity and permeability have the potential to hold the greatest quantities of oil because oil (and other fluids) can travel through them readily.

9. Where is most of the world's oil found?
The majority of oil is found in the Persian Gulf region of the Middle East.

10. How is coal formed?
Coal represents the remains of plants that lived and died in moist or swampy environments, in which there was little bacterially induced decay. As trees and other plants died, their organic matter continued to accumulate, and burial compaction eliminated some of the volatile materials, forming peat from the remains. With additional heat and pressure derived from later, deeper burial under layers of sediment and rock, lignite, bituminous, and ultimately anthracite coal can be produced. Each higher rank in the peat-coal transition is closer to pure carbon in composition.

11. Explain how coal is transformed in coal rank from peat to anthracite coal.
    Heat and pressure during burial burn off volatile matter, which is in greatest abundance in peat and least abundance in anthracite.

12. Describe the two main methods by which coal is mined.
    Near the surface, coal is strip-mined, with overlying strata forcibly removed by a machine called a drag line. Deeply buried coal is acquired through underground mining, in which miners tunnel into the ground and produce a maze of artificial caverns into which coal harvesting machines are introduced.

13. What are some of the environmental drawbacks of mining and burning coal?
    Underground mining introduces the risks of roof collapse, deadly gas, and coal dust. Strip mining potentially destroys soil profiles and natural habitats for wildlife. Much coal has a high sulfur content, and burning of this coal releases sulfur dioxide, which acidifies rain. Acid mine drainage arises when sulfur-rich minerals, such as pyrite, are partly dissolved to produce a dilute sulfuric acid solution that may enter surface streams.

14. Describe how a nuclear reaction is initiated and controlled in a nuclear reactor.
    A neutron is struck against unstable uranium (U-235) within uranium oxide inside a fuel rod, inducing fission of the uranium atom, which emits three additional neutrons which strike other uranium atoms, which release yet more neutrons, forming a chain reaction. The reaction is controlled by keeping the quantity of uranium within the rods below a critical mass and by surrounding the rods with cool, circulating water.

15. Where does uranium form in the Earth's crust? Where does it usually accumulate in minable quantities?
    Uranium is incorporated into magmas, which rise up through the crust, cooling to form granite plutons. Hot groundwater dissolves the uranium scattered throughout the plutons and precipitates it in higher concentrations in veins within the plutons.

16. What are some of the drawbacks of nuclear energy?
    The possibility of meltdown (fuel becoming too hot and melting through some of the architecture of the power plant) as occurred at Chernobyl in the Ukraine, and the difficulties associated with nuclear waste are the primary drawbacks of nuclear energy. Waste must be disposed of in a way that will ensure that radioactive isotopes will not be carried by surrounding groundwater.

17. What is geothermal energy? Why is it not more widely used?
    Geothermal energy is the internal heat of the Earth. There are few places on Earth where hot groundwater rises naturally to sufficiently shallow depths, so that its heat can be readily exploited. In many cases, attempts at utilizing geothermal energy have not proved economically feasible.

18. What is the difference between renewable and nonrenewable resources?

    Renewable resources are those that are naturally replenished at sufficient rates to offset current usage of the resource. Nonrenewable resources include the fossil fuels, which develop over time scales of thousands to millions of years.

19. What is the likely future of oil production and use in the next century?

    Within 50 to 150 years, all of the attainable oil will have been used up (at current rates of usage). As oil becomes scarce in the future, its price will undoubtedly go up and alternative sources of energy will have to be exploited.

Test bank

1. Before the discovery of coal, humans did not produce or consume energy.
   A. true   B. false

2. Which of the following lists contains only fossil fuels?
   A. coal, oil, natural gas   B. coal, geothermal, wind
   C. coal, wood, natural gas   D. hydroelectric, geothermal, wind

3. Which of the following is not one of the five fundamental sources of energy?
   A. nuclear fusion within the Sun
   B. geothermal energy
   C. energy stored within chemical bonds
   D. fossil fuels

4. Wind energy is derived from which pair of fundamental energy sources?
   A. nuclear fission and nuclear fusion
   B. energy within chemical bonds and gravity
   C. nuclear fission and gravity
   D. nuclear fusion and gravity

5. In photosynthesis, plants and algae combine carbon dioxide and water to produce _____.
   A. nitrous oxide and carbon monoxide
   B. methane and ammonia
   C. sugar and sulfur dioxide
   D. sugar and oxygen

6. Tides on the world's oceans are primarily brought about by the gravitation attraction of _____.
   A. the Sun   B. the Moon   C. Jupiter   D. Venus

7. Chemically, oil and gas are both _____.
   A. pure forms of carbon   B. carbohydrates
   C. hydrocarbons   D. carbonate minerals

8. Hydrocarbons consisting of short carbon chains are _____.
    A. more volatile than hydrocarbons with longer chains
    B. more viscous than hydrocarbons with longer chains
    C. typically stable in a solid state at room temperatures
    D. All of the above are correct.

9. Most of the hydrocarbons within oil and natural gas are derived from the breakdown of organic matter from once-living _____.
    A. dinosaurs   B. plankton
    C. terrestrial plants   D. mammals

10. Organic matter builds up evenly over the entire surface of the ocean floor.
    A. true   B. false

11. A buried body of rock that is induced by heat to exude oil is termed a _____.
    A. reservoir rock   B. seal rock   C. source rock

12. A permeable and porous rock, regardless of lithology, is a good candidate to serve as a _____ in an oil-producing scenario.
    A. reservoir rock   B. seal rock   C. source rock

13. An impermeable rock, regardless of lithology, is a good candidate to serve as a _____ in an oil-producing scenario.
    A. reservoir rock   B. seal rock   C. source rock

14. Which organic substance is produced by black organic shales at temperatures below the oil window?
    A. gold   B. coal   C. kerogen   D. keratin

15. Natural gas _____.
    A. consists of longer carbon chains as compared to oil
    B. is denser than oil
    C. is produced at higher temperatures than oil
    D. is produced at higher temperatures than graphite

16. The greatest depth at which oil is found is _____.
    A. 1 km below the surface   B. 6.5 km below the surface
    C. at the Moho   D. the base of the lithosphere

17. Which fossil fuel, oil or natural gas, is found at greater depths within the Earth, and why?
    A. oil, because it is more stable at high temperatures
    B. oil, because it is more stable at low pressures
    C. gas, because it is more stable at high temperatures
    D. gas, because it is more stable at low pressures

18. A black, organic-rich shale could likely serve as either of which two necessary types of rocks within oil fields?
    A. source rock or seal rock
    B. reservoir rock or seal rock
    C. source rock or reservoir rock

19. All porous rocks are highly permeable.
    A. true   B. false

20. Permeability is a measure of _____.
    A. how many pores are present within a rock
    B. the percentage of rock volume that is comprised of pores
    C. how well the pores within rock are connected
    D. the mass of a body of rock divided by the mass of an equal volume of water

21. If a drilled well encounters both oil and natural gas, the _____.
    A. gas will naturally be floating on top of the oil
    B. oil will float above the level of the gas
    C. gas and oil are expected to be thoroughly mixed together
    D. gas will exist in a dissolved state within the oil

22. Shale, salt, and fine-grained limestone that is not fractured are all good candidates to serve as _____ within an oil field.
    A. a reservoir rock   B. a seal rock
    C. a source rock   D. either a source rock or a reservoir rock

23. Oil was first successfully recovered at a town called Titusville in 1859 in which state?
    A. Texas   B. Alaska
    C. California   D. Pennsylvania

24. Exxon, Chevron, Mobil, and Amoco were all once part of a single oil company called _____.
    A. Arco   B. BP   C. Standard Oil   D. Gas-N-Go

25. Oil taken directly from the ground, without artificial chemical alteration, is termed _____.
    A. crude oil   B. raw oil   C. crass oil   D. protooil

26. Oil drilling not only provides gasoline and electric power but also material incorporated in _____.
    A. textiles   B. plastics   C. rubber   D. cement

27. Most of the world's coal was deposited in coal swamps during the _____.
    A. Cretaceous   B. Ordovician
    C. Carboniferous   D. Jurassic

28. Coal is the altered remains of ancient _____.
    A. dinosaurs	B. plankton
    C. terrestrial plants	D. mammals

29. Most coal is mined from _____, which develop and preserve the thick sedimentary sequences necessary for deep burial.
    A. domes	B. basins
    C. shields	D. active margins

30. Which sequence of coal ranks is ordered from lowest to highest?
    A. anthracite, bituminous, lignite
    B. lignite, bituminous, anthracite
    C. bituminous, lignite, anthracite
    D. bituminous, anthracite, lignite

31. Which gas is most abundantly produced from the burning of coal?
    A. sulfur dioxide	B. carbon monoxide
    C. carbon dioxide	D. oxygen

32. At equal volumes, burning which material produces the most heat energy?
    A. anthracite	B. bituminous
    C. lignite	D. peat

33. Coal layers buried under less than 100 m of sediment are usually mined with which technique?
    A. underground mining	B. strip mining

34. Miners used to take canaries into coal mines primarily because _____.
    A. their singing provided companionship
    B. their singing increased productivity
    C. they would die of gas poisoning before the miners would, providing the miners with a warning system
    D. they ferried notes among workers on opposite sides of the mines

35. Nuclear reactions within power plants do not produce the explosive reactions characteristic of atomic bombs primarily because _____.
    A. the isotope of uranium used in power plants never undergoes explosive fission and therefore was never used in bombs
    B. the amount of uranium within nuclear fuel rods is enough to sustain a chain reaction but is less than the critical mass necessary for the reaction to become explosive

36. U-235, the isotope of uranium commonly utilized in nuclear power plants, is _____.
    A. the most common of the naturally occurring isotopes of that element
    B. heavier than the other well-known isotope of uranium
    C. rare even in uranium oxide deposits, and thus usable reactor fuel must be enriched with respect to this isotope

37. Aside from the possibility of a meltdown, nuclear power provides no other major problems or concerns.
    A. true				B. false

38. Geothermal energy provides a low-cost alternative to fossil fuels throughout most of the populated world.
    A. true				B. false

39. Which of the following is a renewable resource?
    A. coal		B. oil		C. wind		D. natural gas

40. The world's supply of oil is likely to be depleted in about _____ years.
    A. ten				B. one hundred
    C. one thousand			D. one million

Chapter 15
Riches in Rocks: Mineral Resources

Learning objectives

1. Students should know a few important precious metal (gold, silver), base metal (copper, lead, iron), and nonmetallic mineral resources (such as dimension stone, quartz sand) and their usefulness.
2. Metals only rarely occur in native, or elemental, form; they are more commonly combined with nonmetals to form ores (mineral compounds that are economically feasible to mine), which must be freed from their crystalline lattice structure (through smelting or other chemical reactions). Oxides, sulfides, and carbonates are commonly ore minerals; silicates usually contain too little metal to be worth pursuing as ore.
3. Humans have combined different metals to form alloys, which may have properties distinct from those of the pure metals combined in the mixture. Bronze, a mixture of copper and tin, was the first widely used alloy and was much more useful in the construction of weaponry and farm implements than was pure copper.
4. Students should know the six types of deposits that concentrate ore minerals (magmatic, hydrothermal, secondary-enrichment, sedimentary, residual mineral, and placer) and how they form.

Chapter summary

Industrial societies use many types of minerals, all of which must be extracted from the upper crust. We distinguish two general categories: metallic resources and nonmetallic resources.

Metals are materials in which atoms are held together by metallic bonds. They are malleable and make good conductors. The use of metals began in prehistoric times.

Metals are derived from ore. An ore is a rock containing native metals or ore minerals (sulfide, oxide, hydroxide, or carbonate minerals with a high proportion of metal) in sufficient quantities to be worth mining. An ore deposit is an accumulation of ore.

Magmatic deposits form when sulfide ore minerals settle to the floor of a magma chamber. In hydrothermal deposits, ore minerals precipitate from hot-water solutions. Secondary-enrichment deposits form when groundwater carries metals away from a preexisting deposit. (Mississippi Valley–type deposits precipitate from groundwater that has traveled long distances through the crust.) Sedimentary deposits, such as banded-iron formations, precipitate out of the ocean. Residual mineral deposits in soil are the result of leaching in tropical climates. Placer deposits develop when heavy metal grains accumulate in coarse sediment along a stream.

Many ore deposits are associated with igneous activity in subduction zones, along mid-ocean ridges, along continental rifts, or at hot spots.

The discovery of ore minerals requires the help of geologists. Economic and environmental concerns govern whether a particular ore deposit can be mined or not. Most mining companies today use open-pit techniques. Mining and processing ore can be a strain on the environment.

Nonmetallic resources include dimension stone for decorative purposes, crushed stone for cement and asphalt production, clay for brick making, sand for glass production, and many others. A large proportion of materials in your home have a geological ancestry.

Mineral resources are nonrenewable. Many are now, or may soon be, in short supply.

Answers to review questions

1. Describe how people have used copper, bronze, and iron throughout history.

Copper was utilized to make arrowheads and scrapers. It was strengthened by the addition of tin to form bronze, which was hard enough to form swords, axes, spears, and plows. Iron is a relatively tough, durable metal. It has been used in cookware and horseshoes, as well as weaponry and ships.

2. Why do most rocks yield little or no useful metals?

Most rocks are composed of silicate minerals, which have a very small proportion of metal.

3. What kinds of concentrations of metal are required for it to be economically minable?

Necessary concentrations are dependent upon the value of the metal and the cost of extraction from the ore. As an example, copper ores that are as little as three-tenths of one percent copper may be economically mined.

4. Describe six different kinds of economic mineral deposits.

Magmatic deposits consist of sulfide ores, which are quite dense. Within magma chambers, they tend to sink to the bottom and crystallize at the floor of the chamber.

Hydrothermal deposits occur when hot groundwater dissolves a metal and reprecipitates it within fractures (to form veins) or pores within rock.

A secondary-enrichment deposit forms, like hydrothermal deposits, because hot groundwater has a tendency to dissolve metals. In this case a primary ore is dissolved, the groundwater travels for some distance, and then precipitates out an enriched, secondary deposit.

Metal ores may be precipitated from seawater and included within layers of sediment at the bottom of the ocean, which ultimately lithify, forming sedimentary deposits. An important example is the banded iron formations which formed during the Proterozoic.

Residual mineral deposits are formed in soils, which develop through extensive leaching of metals by rainwater and concentrated redeposition of the metals in the zone of accumulation.

Placer deposits are derived from the sorting behavior of streams; heavy metallic grains settle out in gravel bars, whereas less dense silicate grains continue to be carried downstream.

5. What procedures are used to locate and mine mineral resources today?
   In settings deemed likely sites for ore deposits, the local strength of Earth's gravitational and magnetic fields is assessed (ores are relatively dense and have high concentrations of magnetic minerals as compared to silicate rock). Rocks, soils, and plants are chemically analyzed to determine if unusually high metal concentrations are present.

6. How is stone cut from a quarry?
   Stone is cut out by hammering with wedges, scraping the rock with abrasive grains, and heating the rock with blowtorch-like cutting jets.

7. Explain the differences between cement, Portland cement, and concrete.
   Cement is a mixture of lime, silica, and oxide minerals with water. As it dries, cement hardens through the precipitation of minerals. Portland cement is a particular cement that is similar in color to the rocks of Portland in England. Concrete is a mixture of cement, sand, and gravel.

8. How many kilograms of mineral resources does the average person in an industrial country use in a year?
   15,000 kg.

9. Compare the estimated lifetimes of ore supplies (in the United States and throughout the world) of iron, aluminum, copper, gold, and chromium.
   Iron: United States, 40 years; world, 120 years.
   Aluminum: United States, 2 years; world, 333 years.
   Copper: United States, 40 years; world, 65 years.
   Gold: United States, 20 years; world, 30 years.
   Chromium: United States, 0 years; world, 75 years.

10. How do political considerations affect the availability of strategic metals?
    Some rare and important metals are predominantly found in countries that are not on friendly terms with the United States. These metals are stockpiled by the government and may be difficult for private citizens to obtain.

11. What are some environmental hazards of large-scale mining?
    Mining creates holes in the surface with associated tailings piles; mine waste may react with water to produce acid runoff. The chemical treatment of ore at smelters also causes pollution.

Test bank

1. Most of the gold prospectors who arrived in California in 1849 were able to amass considerable fortunes.
   A. true   B. false

2. Nearly all of the minerals that are commercially exploited are either native metals or ore minerals.
   A. true   B. false

3. The high electrical conductivity associated with metallic bonds is found in _____.
   A. all minerals that contain metallic ions, in any proportions
   B. minerals that have a sufficient metal content to be considered ores
   C. native metals and pure metals artificially derived from ores
   D. magnetic minerals only

4. The high electrical conductivity associated with metallic bonds results from _____.
   A. the ability of outer electrons to move freely from atom to atom
   B. their radioactive instability
   C. the ability of nuclei to be discharged into the atmosphere
   D. their relatively great atomic mass

5. The ability of a metal to be bent, molded, and stretched is termed _____.
   A. tempering   B. cold working
   C. malleability   D. dexterity

6. The first ores which were widely smelted by humans to produce metal were those of _____.
   A. bronze   B. copper   C. gold   D. iron

7. The first widely used alloy in ancient civilizations was _____.
   A. bronze   B. copper   C. gold   D. iron

8. Because of their economic importance to society, copper and aluminum are considered precious metals.
   A. true   B. false

9. Heating iron ores does not produce native iron unless done in the presence of _____.
   A. oxygen   B. carbon dioxide
   C. nitrous oxide   D. methane

10. Most of the metals currently in commercial use _____.
    A. were known to the ancients
    B. are found only in North America and Asia
    C. were discovered during the last two centuries
    D. are excavated as native metals and do not require smelting

11. Any mineral that contains a high proportion of metallic ions is likely to be a suitable ore mineral.
    A. true   B. false

12. Which of the following is not an important ore of iron?
    A. magnetite (iron oxide)
    B. hematite (iron oxide)
    C. pyrite (iron sulfide)
    D. All of the above minerals are important ores of iron.

13. Which type of ores are most commonly concentrated in magmatic deposits?
    A. oxides
    B. sulfides
    C. native metals
    D. copper ores

14. Which property of these ores (answer to question #13) causes them to be concentrated by magmas?
    A. low temperatures of crystallization
    B. high temperatures of crystallization
    C. low density
    D. high density

15. Mineral-rich veins within plutons, deposited by hot groundwater into fractures within the rock, characterize _____.
    A. hydrothermal deposits
    B. placer deposits
    C. residual mineral deposits
    D. sedimentary deposits

16. Secondary-enrichment deposits are most similar in mode of formation to which other deposit type?
    A. hydrothermal deposits
    B. placer deposits
    C. residual mineral deposits
    D. sedimentary deposits

17. Which ore minerals are commonly found in ancient sedimentary deposits that are two billion years old?
    A. copper sulfides
    B. aluminum oxides
    C. iron oxides
    D. copper oxides

18. The stereotypical gold rush prospector panning for gold in a stream bed is an example of exploiting _____.
    A. magmatic deposits
    B. placer deposits
    C. residual mineral deposits
    D. sedimentary deposits

19. Ores that are found in veins within plutons typically occur _____.
    A. by themselves as pure mineral assemblages
    B. together with other ores, but with no other minerals
    C. alongside abundant calcite
    D. alongside abundant quartz

20. The blocks of rock used in construction are termed _____.
    A. dimension stone
    B. gemstone
    C. ore
    D. lodestone

21. Open pits at the surface of the Earth, constructed to extract stone for building purposes are called _____.
    A. open pit mines
    B. underground mines
    C. quarries
    D. sedimentary deposits

22. Because it can be split easily into thin sheets and is impermeable, _____ is commonly used to make roofing tiles.
    A. limestone
    B. sandstone
    C. gneiss
    D. slate

23. Concrete is _____.
    A. synonymous with the term "cement"
    B. synonymous with the term "Portland cement"
    C. a combination of cement, sand, and gravel
    D. powdered rock of any lithology mixed with water and allowed to set overnight

24. Stone is typically broken into slabs and blocks through the use of _____.
    A. mechanical splitting and chiseling
    B. abrasion
    C. heat
    D. All of the above techniques are used.

25. Which mineral resources are considered renewable?
    A. base metals only
    B. nonmetallic minerals only
    C. iron and aluminum ores
    D. No mineral resources are renewable.

26. The United States has active mines within its boundaries that are sufficient to maintain a steady supply of all strategically important metals.
    A. true   B. false

27. As compared to coal mining, there is virtually no adverse environmental impact of mineral mining.
    A. true   B. false

28. Commercial ore deposits are most likely to be found associated with _____.
    A. thick basinal sandstones and shales
    B. blueschists
    C. evaporite sequences
    D. igneous rocks

29. Most important mineral resources, including ores, can be expected to last for the next one thousand years.
    A. true   B. false

30. The presence of gold in western South America is unrelated to the subduction zone nearby.
    A. true   B. false

Chapter 16
Unsafe Ground: Landslides and Other Mass Movements

Learning objectives

1. Students should be able to differentiate the major types of subaerial (slump, slide, flow, lahar, avalanche) and submarine (creep, slump, flow, turbidity current) mass movements on the basis of coherence, geometry, velocity, and other physical characteristics.
2. They should be aware of the major factors that favor mass movement: steep slopes, addition of mass upslope, excavation downslope, supersaturation of pore space, barren hillsides, and active seismicity.

Answers to review questions

1. What four factors distinguish the various types of mass movement?
    Mass movements are distinguished on the basis of the material that is moving, its rate of movement, the physical character of the moving mass (rigidity and viscosity), and the environment (subaerial or submarine).

2. How does a slump differ from creep? How does it differ from a mudflow or debris flow?
    Creep is gradual; a slump is an event in which a mass of material breaks off of an elevated slope and slides downward along a curved surface. Mudflows and debris flows contain more abundant water in the moving mass than does a slump.

3. How does a rock or debris slide differ from a slump? How does it differ from a mud- or debris flow?
    Slumps detach and slide along curved (spoon-shaped) surfaces; rock and debris slides travel down planes that are subparallel to the slope of the hillside. Rock and debris slides are falls of dry rock and sediment; mud- and debris flows involve mixtures of sediment with water.

4. How are submarine slumps similar to those above water? How are they different?
    Both types of slump consist of the detachment and sliding of a semi-coherent mass of sediment. In the submarine environment, however, the slump sediment (and thus the history of the event) is typically preserved due to eventual burial by overlying sediments.

5. What factors cause the relief of the Earth's surface? How does relief affect mass movements?
    Tectonic activity at convergent margins and collision zones thicken the crust and produce uplift. Normal faulting in continental rifts leaves footwall mountain blocks elevated with respect to the hanging wall. Changes in the temperature of the mantle may cause broad regions to be uplifted or downwarped. Volcanism produces elevated mountains at hot spots, subduction zones, and rifts. Deposition of sediment commonly produces mounds or dunes.

Humans may construct mounds or channels, as does running water. Seashore slopes may arise when seawater becomes bound up in continental glaciers.

All of these factors produce local slopes; the steeper the slope, the faster a given episode of mass movement will move.

6. How does a small amount of water between grains hold material together? How does this change when the sediment is oversaturated?

Slightly wet sediment is held together because of surface tension: the polarity of water molecules provides attractive force among the molecules and between them and mineral surfaces. However, at oversaturation, the pores between grains become filled with water under pressure, which pushes grains apart.

7. What force is responsible for downslope movement? What force helps resist that movement?

Gravity initiates downslope movement; it is opposed by friction, chemical bonds within and among minerals, mineral cement, and surface tension.

8. How does the angle of repose change with grain size? How does it change with water content?

Larger (especially irregularly shaped) grains can take on a higher angle of repose. A slight amount of water within sandy sediment will increase the angle of repose until the water content becomes too great and leads to instability.

9. What factors trigger downslope movement?

Vibrations (as from seismic waves), increasing slopes, increasing the load upslope, and excavating the hillside downslope may trigger mass movement downslope.

10. What factors contribute to the high frequency of landslides in coastal southern California? Why don't people just avoid building in these dangerous areas?

Southern California is dominated by a transform plate boundary. Faulting has produced highly fractured rock conducive to chemical weathering, resulting in loose, unstable substrates in a region with frequent earthquakes. Many Californians build atop cliffs that are subject to wave erosion. Rain is infrequent in southern California, so vegetation is scarce, but rain may occasionally pour down in torrents. Nevertheless, the warm weather and social and recreational opportunities make southern California a popular place to live. Oceanfront property has an especially strong appeal (you only have neighbors on three sides).

11. How do geologists predict whether an area is prone to mass wasting?

Locally, geologists assess the risk of mass movement by looking for sedimentary evidence of past events and bends in tree trunks and trying to determine the age of exposed surfaces. Cracking in manmade structures may indicate the coming onset of mass movement. Over broader regions, geologists assess slope angle, sediment coherence, bedrock lithology, vegetation, climate, water saturation, fractures, seismic risk, and the potential that the slope may be undercut by waves or currents.

12. What steps can people take to reduce the risk of mass wasting?
Adding vegetation, reducing slope, preventing undercutting by diverting rivers or breaking waves offshore, and developing structures that artificially stabilize slopes are methods that can be taken to reduce the risk of mass movements.

Test bank

1. Which of the following types of mass movement takes place most gradually?
   A. slump   B. creep   C. rock slide   D. mudflow

2. Which of the following types of mass movement is least coherent (most like a liquid)?
   A. slump   B. creep   C. rock slide   D. mudflow

3. Which of the following types of mass movement travels down a curved surface?
   A. slump   B. creep   C. rock slide   D. mudflow

4. Strictly speaking, which of the following types of mass movement is a landslide?
   A. slump   B. creep   C. rock slide   D. mudflow

5. Where would you expect to see solifluction?
   A. Mexico   B. Italy   C. Egypt   D. Siberia

6. Slumps typically arrive at the bottom of a slope intact.
   A. true   B. false

7. The principal difference between a debris flow and a mudflow is the _____.
   A. shape of the path taken by the moving mass
   B. grain size of the moving mass
   C. former mass contains abundant water whereas the latter is dry
   D. former mass contains pyroclastic debris from a volcanic eruption

8. The principal difference between a lahar and a mudflow is the _____.
   A. shape of the path taken by the moving mass
   B. grain size of the moving mass
   C. former mass contains abundant water whereas the latter is dry
   D. former mass contains pyroclastic debris from a volcanic eruption

9. The principal difference between a debris flow and a debris slide is the _____.
   A. shape of the path taken by the moving mass
   B. grain size of the moving mass
   C. former mass contains abundant water whereas the latter is dry
   D. former mass contains pyroclastic debris from a volcanic eruption

10. The principal difference between a slump and a debris slide is the _____.
    A. shape of the path taken by the moving mass
    B. grain size of the moving mass
    C. former mass contains abundant water whereas the latter is dry
    D. former mass contains pyroclastic debris from a volcanic eruption

11. The principal difference between a debris slide and a debris avalanche is the _____.
    A. shape of the path taken by the moving mass
    B. grain size of the moving mass
    C. former mass contains abundant water whereas the latter is dry
    D. former mass contains pyroclastic debris from a volcanic eruption

12. Of subaerial and submarine mass movements, we have a much better historical record of _____.
    A. subaerial movements   B. submarine movements

13. Which of the following types of mass movement is likely to move most rapidly?
    A. slump   B. creep
    C. rock slide   D. debris slide

14. In which list of submarine mass movements are the events ordered in terms of the proportion of water included in the moving mass (least water to most water)?
    A. turbidity current, submarine debris flow, submarine slump
    B. submarine debris flow, submarine slump, turbidity current
    C. submarine slump, submarine debris flow, turbidity current
    D. submarine slump, turbidity current, submarine debris flow

15. The immediate cause of incidents of mass movement is _____.
    A. electromagnetic attraction
    B. gravitation
    C. magnetism
    D. friction

16. A hillside consisting of a pile of sediment in which the slope is more gentle than the angle of repose for the sediment represents a(n) _____.
    A. stable slope   B. unstable slope

17. Which of the following factors decreases the risk of mass movement?
    A. nearby earthquakes
    B. excavation into the base of a hill
    C. adding weight to the top of a hill
    D. adding vegetation to the side of a hill

18. Which type of bedrock provides the greatest risk of landslide?
    A. granite   B. gneiss   C. sandstone   D. shale

19. Which of the following increases the risk of mass movement?
    A. adding a small amount of moisture to loose, dry sediment
    B. waves breaking before they reach sea cliffs
    C. flooding the sediment with water beyond saturation
    D. reducing the grade of the slope

20. Mass movement along coastal cliffs in southern California is so common because _____.
    A. the region is along an active margin with frequent earthquakes
    B. the region is dry and the slopes support little or no vegetation
    C. much of the rock in southern California is fractured
    D. All of the above.

Chapter 17
Streams and Floods: The Geology of Running Water

Learning objectives

1. Students should be familiar with stream channels and should be able to label a diagram with point bars, cut banks, oxbow lakes, meander necks, and the thalweg. They should know how meanders form and how they reinforce themselves, as well as the relationship among erosion, deposition, and current velocity. They should know geometrical differences between braided and meandering streams and how streams become braided.

2. They should be able to identify the four major drainage networks (dendritic, radial, rectangular, and trellis) and understand the conditions that give rise to each.

3. Rivers carry a sediment load in three components: dissolved ions, suspended grains (the greatest portion volumetrically), and bed load (larger grains which tumble along the stream bottom). Stream competence refers to the largest particle size a stream can carry; the total volume of sediment carried by a stream is its capacity.

4. The total volume of water passing by a fixed point in one second is its discharge; the Amazon has by far the greatest discharge of all the rivers on Earth. In moist climates, discharge increases downstream with the addition of tributaries and rainfall; however, in arid climates, discharge may decrease downstream because of evaporation (and human usage in agriculture).

5. The farthest that a stream can downcut into underlying sediment or rock is termed the base level. Sea level is the ultimate base level, but resistant rock and lakes may locally raise the base level substantially.

6. Streams are important agents of erosion, carving out steep-walled canyons and more shallow-walled valleys. Water itself is not especially abrasive, but the sediment that it carries is highly so.

7. When rivers meet the sea (or some other still body of water), their velocity and competence decrease, and they deposit sediments, which form wedge-shaped bodies termed deltas. The shapes of ocean deltas are controlled by sediment supply, the strength of ocean waves and currents, and the tides.

8. Floods may be seasonal or episodic; they may arise suddenly or gradually (in the floodplains of major river systems). Large, catastrophic floods are less common than small floods, and so floods of various magnitudes have been assigned estimated recurrence intervals (average waiting times between floods of a specific magnitude over very long time periods). For example, a flood with a 100-year occurrence interval has a 1% chance of occurring within a given year.

Answers to review questions

1. Why are continental divides often near active continental margins?
    Active margins have mountain ranges which impede the flow of water in the direction of the active margin on the landward side of the range.

2. What is the continental divide in North America? Where does it run?

The continental divide separates streams and rivers that lead to the Pacific Ocean from those that ultimately drain into the Gulf of Mexico, part of the Atlantic Ocean; it runs north-south through the Rocky Mountains.

3. Describe the four different types of drainage networks. What factors are responsible for the formation of each?

Dendritic drainage networks appear similar to the branching patterns of vegetation and occur when streams cut through uniform material with a relatively constant slope. Radial drainage resembles the spokes of a wheel and occurs when streams drain a central, conical mountain. Rectangular drainage consists of streams which make right-angle turns, with multiple streams commonly traveling in parallel formation; they occur in areas where perpendicular sets of joints present paths of least resistance to flow. Trellis drainage occurs in valley and ridge topography, with major streams inhabiting the valleys and traveling parallel to the ridges, and minor streams draining the hillsides transversely before their flows join the valley streams.

4. How are superposed and antecedent drainages similar? How are they different?

Superposed and antecedent streams are similar in that their courses are independent of the attitude of material into which they cut. Superposed streams occur in flat-lying areas, cutting down into previously buried folded rocks. Antecedent streams cut through mountain ranges, which these streams predate.

5. What factors determine whether a stream is permanent or ephemeral?

In a moist climate, streams are permanent because the water table in nearby sediment is above the level of the stream floor; in a dry climate, the water table sinks lower and streams may only flow after snowmelt or a torrential rain, succumbing off and on to evaporation.

6. How does discharge vary according to the stream's length, climate, and position along the stream course?

In a moist climate, discharge increases steadily due to the addition of water from tributaries and rainfall on the journey to the sea. In a dry climate, discharge may decrease downstream as the stream is subjected to intense evaporation and relatively little rainfall.

7. Why is average downstream velocity always less than maximum downstream velocity?

At the contact with sediment, water encounters friction and moves more slowly than at the surface near the middle of the stream.

8. How does a turbulent flow differ from a laminar flow?

In laminar flow, all parcels of water travel in the same direction. Turbulent flow is chaotic, featuring the development of circular eddy currents.

9. Describe how streams and running water erode the Earth.
   Running water can break and lift particles off the stream bottom and also use these particles as abrasive tools to scratch away additional sediment. Water can dissolve some minerals within the stream bed.

10. What are three components of sediment load in a stream?
    Dissolved load (ions), suspended load (small, floating grains), and bed load (larger grains moving along the stream bottom).

11. Distinguish between a stream's competence and its capacity.
    Competence is a measure of stream energy and refers to the largest particle size that a stream can carry; capacity is the amount of sediment that a stream is carrying.

12. Why is the average stream velocity actually slower in a turbulent mountain stream than in a broad flowing river downstream?
    Turbulent flow involves swirling eddy currents, which flow counter to the overall average direction of flow and oppose this flow. Downstream, flow is smooth and laminar.

13. What factors determine the position of the base level?
    The local base level of a stream may be more elevated than it would otherwise be if lakes, resistant rock layers, and large streams are present nearby.

14. What do lakes, rapids, and waterfalls indicate about the stream gradient and base level?
    Lakes indicate a shallow gradient and a relatively high base level; rapids and waterfalls are indicative of high stream gradients, where the base level is well below the stream bottom (aside from the region of resistant rock at the head of a waterfall where the base level may be elevated).

15. How does a braided stream differ from a meandering stream?
    Braided streams are choked with relatively coarse sediment; they readily spill through their banks to form numerous, subparallel braided channel strands. A meandering stream consists of a single, highly sinuous channel (which may, however, change course over time).

16. Describe how meanders form, develop, are cut off, and then abandoned.
    All streams will waver from a straight line at some point. When they do, the outer bend of a turn has relatively fast flow (becoming a site of erosion) as compared to the inner bend (which becomes a site of point bar deposition). Meanders are thus a system of positive feedback, as subsequent erosion and deposition push the stream channel farther and farther off a straight course. However, streams change their course over time and can erode through their banks to cut off the meander with a shorter, straighter segment which draws flow away from the meander and isolates it, forming an oxbow lake.

17. Describe how deltas grow and develop.
    Rivers carrying sediment meet an ocean or large lake, in which the water is stagnant. Flow velocity is slowed dramatically at the interface, and the stream loses its competence, its sediment load falling out to form a wedge.

18. What is stream piracy?
    Stream piracy occurs when two channels come to intersect, with the stream designated as the pirate stealing flow from the stream designated as the victim.

19. What human activities tend to increase flood risk and damage?
    Living and working within the floodplain of major rivers bring increased risk; additionally, buildup of the "asphalt jungle" makes flooding more common as it makes it harder for rainwater to infiltrate into the sediment.

20. What is the recurrence interval of a flood? Why can't someone say that "the hundred-year flood happened last year, so I'm safe for another hundred years?"
    The occurrence interval of a flood is the average waiting time between successive floods of a given magnitude (or greater). Flooding behavior is unpredictable, and successive flooding events of a given magnitude cannot be expected to be evenly spaced.

21. How have humans abused and overused the resource of running water?
    Through pollution, overzealous damming, and agricultural irrigation in desert areas.

Test bank

1. The flat-lying area surrounding a river channel is termed the _____.
    A. base level   B. floodplain
    C. stream gradient   D. thalweg

2. The lowest elevation to which a stream can downcut is the _____.
    A. base level   B. floodplain
    C. stream gradient   D. thalweg

3. The middle part of the channel, where the stream flows fastest, is the _____.
    A. base level   B. floodplain
    C. stream gradient   D. thalweg

4. The slope of a streambed measured at some point along its course is the _____.
    A. base level   B. floodplain
    C. stream gradient   D. thalweg

5. Continental divides are commonly found near passive margins.
    A. true   B. false

6. In uniform sediments with a relatively constant slope at the surface, a _____ drainage network is expected.
    A. dendritic   B. radial
    C. rectangular   D. trellis

7. In a region with prominent orthogonal sets of joints, a _____ drainage network is expected.
    A. dendritic   B. radial
    C. rectangular   D. trellis

8. In a region characterized by a parallel series of ridges and valleys, a _____ drainage network is expected.
    A. dendritic   B. radial
    C. rectangular   D. trellis

9. In the region immediately surrounding an isolated volcano, a _____ drainage network is expected.
    A. dendritic   B. radial
    C. rectangular   D. trellis

10. The course of rivers always avoids passing through mountain ranges.
    A. true   B. false

11. If a stream cuts through flat-lying sediments to reach an anticline produced in an ancient orogenic event, the course of the stream will likely be unaffected.
    A. true   B. false

12. The water table typically lies above the bottom of a stream in a _____ climate.
    A. hot   B. cold   C. moist   D. dry

13. Ephemeral streams _____.
    A. consist of a series of intertwined channels that are overloaded with sediment
    B. have flowing water either episodically or during a portion of the year
    C. have a main channel that is highly sinuous (curvy)
    D. are those that "steal" water from other streams, which they intersect

14. Meandering streams _____.
    A. consist of a series of intertwined channels that are overloaded with sediment
    B. have flowing water either episodically or during a portion of the year
    C. have a main channel that is highly sinuous (curvy)
    D. are those that "steal" water from other streams, which they intersect

15. Pirate streams _____.
    A. consist of a series of intertwined channels that are overloaded with sediment
    B. have flowing water either episodically or during a portion of the year
    C. have a main channel that is highly sinuous (curvy)
    D. are those that "steal" water from other streams, which they intersect

16. Braided streams _____.
    A. consist of a series of intertwined channels that are overloaded with sediment
    B. have flowing water either episodically or during a portion of the year
    C. have a main channel that is highly sinuous (curvy)
    D. are those that "steal" water from other streams, which they intersect

17. Which has greater competence?
    A. a small mountain stream   B. the Mississippi River

18. Which has greater capacity?
    A. a small mountain stream   B. the Mississippi River

19. Which of these rivers has the greatest volumetric discharge?
    A. the Mississippi   B. the Nile
    C. the Amazon   D. the Congo

20. The average velocity of a stream is always greatest near its source.
    A. true   B. false

21. All else being equal, a deep, narrow stream will flow _____ a shallow, broad stream.
    A. more slowly than
    B. more rapidly than
    C. at the same rate as

22. Moving away from its headwaters, the flow of a stream becomes more _____.
    A. turbulent   B. chaotic
    C. competent   D. laminar

23. The sediment load of a stream consists of only those grains that are fine enough to stay in suspension.
    A. true   B. false

24. In the Mississippi River, most of the sediment load transported to the Gulf of Mexico _____.
    A. bounces along the river bed
    B. is carried as suspended grains
    C. travels in the form of dissolved ions

25. As the velocity of flow decreases, _____.
    A. flow tends to become more laminar
    B. the maximum flow velocity decreases
    C. suspended sediment starts to be deposited
    D. All of the above are correct.

26. Ultimately, the base level of a stream valley can be no lower than _____.
    A. the average elevation of the continent on which it is found
    B. sea level
    C. the average elevation of the ocean basins

27. The base level of a stream rises when it encounters a lake because _____.
    A. rivers cannot downcut as rapidly as lakes can
    B. contacting the lake causes the flow of a stream to slow down

28. The distinction between a valley and a canyon is _____.
    A. nonexistent; the two terms are synonymous
    B. the sides of a canyon are more steep than those of a valley
    C. the sides of a valley are more steep than those of a canyon
    D. valleys are often cut by streams; canyons are eroded out by the wind

29. At a waterfall, where two sedimentary lithologies are juxtaposed, which lithology is more likely to form the cliff over which the water falls?
    A. sandstone   B. shale

30. Within a meander, where is sediment most likely to be deposited?
    A. on the outer banks of the meander
    B. on the inner banks of the meander
    C. uniformly to either side of the meander

31. The inner edge of a meander, where sediment is deposited, is a(n) _____.
    A. meander neck   B. cut bank
    C. point bar   D. abandoned meander

32. The outer edge of a meander, where material is being eroded, is a(n) _____.
    A. meander neck   B. cut bank
    C. point bar   D. abandoned meander

33. A meander that is cut off to become completely isolated from the main channel, and which dries up, is a(n) _____.
    A. meander neck   B. cut bank
    C. oxbow lake   D. abandoned meander

34. A meander that is cut off to become completely isolated from the main channel, but which retains water, is a(n) _____.
    A. meander neck   B. cut bank
    C. oxbow lake   D. abandoned meander

35. A thin strip of land separating abutted cutbanks on opposite ends of a meander loop is a(n) _____.
    A. meander neck   B. cut bank
    C. point bar   D. abandoned meander

36. The deltas of all major rivers consist of multiple radiating lobes of sediment in a "bird's foot" configuration, similar to the Mississippi River delta.
    A. true   B. false

37. Sediment buildup on either side of a stream channel produces _____.
    A. meander necks   B. cut banks
    C. natural levees   D. abandoned meanders

38. All flooding events occur so rapidly that there is no time to alert people in harm's way.
    A. true   B. false

39. A 100-year flood is more catastrophic than a 50-year flood.
    A. true   B. false

40. If a 50-year flood occurs on the Mississippi River in 2003, what is the probability that a flood of the same magnitude will occur in 2004?
    A. zero   B. 1%   C. 2%   D. 50%

Chapter 18
Restless Realm: Oceans and Coasts

Learning objectives

1. Students should be broadly familiar with the bathymetric profile of the world ocean and be able to picture the depths, slope, and extent of continental shelves, slopes, rises, and the abyssal plains (punctuated by the mid-ocean ridges).

2. They should know that ocean circulation is dominated by two factors (temperature and salinity). The range of temperatures in the world ocean is more extensive than the range of salinity, so modern circulation is dominated by contrasts in temperature, with cold water sinking at the poles and creeping toward the equator along the bottom. In warm restricted seas, circulation can be induced by the evaporation of surface water to produce denser, more saline water. The freezing of water at the poles increases salinity, and rainfall in the tropics reduces it.

3. Surface currents, though generated by winds, do not flow in parallel with the wind due to the Coriolis effect. Water currents in the northern and southern oceans are dominated by clockwise and counterclockwise circular gyres, respectively. Where currents primarily flow towards the land, downwelling results; where currents flow away from the coastline, upwelling develops.

4. Tides are primarily generated by the Moon's gravitational pull on the Earth. The attraction is strongest on the side of the Earth facing the Moon, and weakest away from the Moon. The Earth is stretched along the Earth-Moon axis, with a sublunar bulge tracking the Moon and a secondary bulge directly opposite to it. Along these bulges, tides are high; the tide is low away from the bulges. The extent of the tidal reach (difference in local sea level at high and low tides) is affected by the position of the Sun (with stronger [spring] tides occurring during full and new moons).

5. Waves affect the water to a depth of one-half their wavelength. When oblique waves reach the shore, the portion of the wave crest nearest the shore "feels the bottom" first and slows down. Thus the wave is refracted (bent) and typically impacts the shore at an angle of less than five degrees. Backwash is gravity driven, falling straight back away from shore. Thus if waves are primarily impacting from a single directional angle (away from head on), the resultant of landward swash and seaward backwash is lateral (longshore) current and associated beach drift.

6. Most human attempts to combat beach drift and other forms of beach erosion have been futile. Groins increase sand deposition in the region directly upcurrent but increase erosion rates in the region directly downcurrent. Breakwaters put in place to quiet the waters in a harbor lead to greater deposition of sand, displacing water.

Answers to review questions

1. How much of the Earth's surface is covered by oceans? How much of the world's population lives within 100 km of a coast?
    70.8%; >60%.

2. Describe the typical topography of a passive continental margin, from the shoreline to the abyssal plain.
    Outward from the shoreline, the continental shelf floor dips very shallowly (less than one-third of 1° from horizontal); the shelf extends horizontally from the coast for a few hundred kilometers, with water depth reaching a few hundred meters. From the shelf edge, the continental slope angles more steeply downward (2°), plunging to a depth of approximately 4 km. From there slope decreases to 0.5° in a region called the continental rise, which extends to a depth of 4.5 km. Beyond this depth, the topography of the ocean is flat abyssal plain.

3. How do the shelf and slope of an active continental margin differ from those of a passive margin?
    At active margins, the shelf is very thin, and the slope dips almost twice as steeply.

4. Where does the salt in the ocean come from?
    Salt and other ions in the ocean are carried in by rivers and groundwater, originating from the chemical weathering of terrestrial rock. Ions are also derived from dissolved volcanic gases. The concentration of salt is greater in the ocean than in rivers because of evaporation.

5. How does the salinity in the ocean vary?
    Salinity is greater in polar regions, which are subjected to freezing, and in dry climates and restricted seaways, due to evaporation. Salinity is below average in tropical areas of high moisture and near the mouths of large rivers. Salinity averages 3.5% and ranges from 1 to 4.1%.

6. How does the temperature of the ocean vary?
    At the poles, ocean water is near freezing (0° C); in restricted tropical seas, the water temperature may be as great as 35° C.

7. How does the density of seawater vary?
    Density increases with decreasing temperature (at least down to 4° C) and with increasing salinity. In the modern ocean, temperature variation is the predominant factor; seawater is densest at the poles.

8. What factors control the direction of surface currents in the ocean?
    Surface currents are controlled primarily by winds and by the Coriolis effect.

9. What is the Coriolis effect, and how does it affect oceanic circulation?

The Coriolis effect is the tendency of projectiles (and ocean currents) to deviate from straight north-south trajectories due to the unequal rotational speed of the Earth (which spins fastest at the equator and slowest at the poles). In the Northern Hemisphere, southbound currents are deflected to the west (as they originate nearer to the pole, where the surface rotates more slowly than at the equator), and northbound currents are deflected eastward. The Coriolis effect deflects currents from the direction of the blowing wind which, through shear, initiates the currents.

10. What is the Ekman spiral, and how does if affect oceanic circulation?

Just as the wind shears the uppermost surface of the ocean, this layer of water shears the water beneath it, which is deflected off the course of the uppermost layer by the Coriolis effect. This process continues with depth, with each lower layer deflected from the course of the layer which shears it from above, forming a downward (Ekman) spiral.

11. How do currents determine where there is oceanic upwelling or downwelling?

Onshore currents create an overabundance of nearshore water, which is alleviated through downwelling. If currents head offshore, the exiting nearshore water is replaced by upwelling from depth.

12. What causes the tides?

Tides are generated by the gravitational attraction of the Moon, moderated by the gravitational attraction of the Sun, in combination with the rotation of the Earth. At the surface facing the Moon, the sea bulges outward toward the Moon because the Moon's gravitational pull is strongest at this closest point. Directly opposite this bulge, another outward bulge arises, because at this most distant point, the Moon's force is weakest. Water responds to the centrifugal effect imposed by Earth's rotation, so it tends to bulge outward; on the side of the Earth farthest from the Moon, this inertial effect is stronger than the Moon's pull. Tides are high in the regions of the bulges, and low opposite them.

13. Describe the motion of water molecules in a wave.

Water molecules in waves on the open ocean move in roughly circular paths (up, forward, down, back).

14. How does wave refraction cause longshore currents?

Wave refraction by itself does not cause longshore currents (see answer to question #17), but aggravates erosion rates in headlands (jutting points of land) by concentrating wave energy there and directing longshore currents into the interior of bays.

15. What is a rip current?

Rip currents are localized, fast currents that head offshore from a beach.

16. How does beach sand migrate as a result of longshore currents?
   Waves are incident upon the beach at angles that vary from the orientation of the coastline by as much as about 5°, but the gravity-driven backwash heads straight outward. Waves pull sand with them, so the net effect of this variance is longshore currents and longshore drift of sand as well.

17. Describe how waves affect a rocky coast.
   Storm waves cause pebbles and boulders (derived from the sea cliffs) to collide, breaking them into finer pieces, and abrade into the cliff faces as well.

18. What is an estuary? Why is it such a delicate ecosystem?
   An estuary is a bay in which seawater and freshwater mix, formed when sea level rises to flood a river valley. Estuaries are isolated pockets of brackish (characterized by intermediate and variable salinity) waters between the freshwater of rivers and the saltwater of the sea; they support a unique ecosystem distinct from that seen in the vast ocean of normal salinity. Toxins introduced from the continent by rivers may build up to dangerous levels in estuaries as they can become bound up by flocculating clay.

19. Describe how a reef system on a seamount evolves from a fringing reef to an atoll.
   Algae within the digestive tissue of reef coral require sunlight, so they only grow in shallow, clear (and warm) seawater. In the middle of deep ocean, a small island provides a shallow-water setting around its perimeter, so a fringing reef develops. As the island subsides, the coral continues to grow upward and stay in shallow water. When the island submerges beneath the surface of the ocean, it is termed a seamount, and the reef now forms an atoll.

20. How do plate tectonics, sea-level changes, sediment supply, and climate change affect the shape of a coastline?
   Coastlines along active margins have relatively steep drop-offs; in most regions that have long been passive margins, broad, flat continental shelves offshore merge into broad, flat coastal plains onshore. Emergent coastlines build elevated terraces of sand which once formed a beach front; submergence of the coastline may produce an irregular coastline shape (as when an estuary rises up and the shoreline assumes the dendritic shape of a river-cut valley). For beaches to build up, the supply of sediment must be greater than the rate of erosion induced by waves. Stormy weather increases erosion rates, so the growth of a beach requires yet greater sediment supply.

21. How does a sea-level change affect human activities in coastal regions?
   Coastal inundation forces humans to move their habitations toward the interior.

22. How does human interference with beach drift cause problems?
   Groins erected to curtail longshore beach drift actually increase the erosion rate of the beach on the lee side of the groin. Breakwaters erected to calm the water behind them cause sand to build up, forcing the water to retreat laterally.

Test bank

1. Most of our knowledge of the topography of the ocean floor was obtained during the _____.
    A. 18th century
    B. 19th century
    C. first half of the 20th century
    D. second half of the 20th century

2. The form and topography (depth profile) of the ocean floor is termed _____.
    A. orogeny   B. bathymetry
    C. allocthony   D. aquitopography

3. Approximately what percentage of the Earth's surface is covered with water?
    A. 30%   B. 50%   C. 70%   D. 85%

4. The two most abundant rocks within oceanic crust are _____.
    A. granite and gabbro   B. basalt and gabbro
    C. granite and rhyolite   D. limestone and shale

5. Compared to continental lithosphere, oceanic lithosphere is _____.
    A. thicker
    B. thinner
    C. approximately the same thickness

6. Compared to continental crust, oceanic crust is _____.
    A. denser   B. less dense   C. about the same density

7. The major oceans, such as the Atlantic, have been present since the Precambrian.
    A. true   B. false

8. The shallowest portion of the ocean is found along the margins of continents in regions termed _____.
    A. abyssal plains   B. continental rises
    C. continental shelves   D. continental slopes

9. Of the choices below, which portions of the seafloor are most steeply tilted?
    A. abyssal plains   B. continental rises
    C. continental shelves   D. continental slopes

10. _____ make up a majority of the ocean floor worldwide.
    A. Abyssal plains   B. Continental rises
    C. Continental shelves   D. Continental slopes

11. The average depth of the abyssal plains below sea level is about _____.
    A. 1 km   B. 2.5 km   C. 4.5 km   D. 8 km

12. The deepest segments of the ocean floor are found _____.
    A. along mid-ocean ridges
    B. in the geographic centers of abyssal plains
    C. in trenches associated with passive margins
    D. in trenches associated with subduction zones

13. Most of the igneous rocks within oceanic abyssal plains _____.
    A. exposed at the surface of the sea floor
    B. covered by sand and gravel introduced by rivers at deltas
    C. covered by clay and the skeletal remains of microplankton
    D. covered by limestones made up of the fragments of large invertebrate shells

14. Submarine canyons are _____.
    A. completely the result of ancient river erosion at times when sea level was lower than at present
    B. deepened by the activity of turbidity currents
    C. exclusively created by turbidity current flow and bear no relation to the courses of terrestrial rivers

15. Turbidity currents give rise to sedimentary deposits called turbidites, which are characterized by _____.
    A. finely laminated sandstones
    B. graded bedding
    C. limestone composed of shell fragments in a muddy matrix
    D. very well sorted, massive sandstone

16. Compared to freshwater, saline ocean water is _____ and therefore provides _____ support to floating objects.
    A. less dense; more buoyant
    B. less dense; less buoyant
    C. more dense; more buoyant
    D. more dense; less buoyant

17. What percentage of seawater, on average, consists of dissolved salt ions?
    A. 1.5%   B. 3.5%   C. 5.5%   D. 6.5%

18. Above 4° C, the density of seawater increases with _____.
    A. increasing temperature and increasing salinity
    B. decreasing temperature and increasing salinity
    C. increasing temperature and decreasing salinity
    D. decreasing temperature and decreasing salinity

19. Most of the dissolved cations in seawater (sodium, potassium, calcium, magnesium) are _____.
    A. introduced by volcanic gases coming into solution
    B. emitted at hydrothermal vents in the deep sea
    C. introduced by the inflow of rivers
    D. derived by the dissolution of salt deposits on the sea floor

20. The densest ocean water forms _____.
    A. around the mouths of larger rivers
    B. in moist, tropical regions
    C. at the poles
    D. above mid-ocean ridges

21. As compared to continental interiors, surface temperatures in coastal regions _____.
    A. are warmer throughout the year
    B. are colder throughout the year
    C. experience a more extreme range throughout the year
    D. experience a less extreme range throughout the year

22. Major oceanic surface currents travel _____.
    A. parallel to the prevailing wind direction in the region
    B. in direct opposition to prevailing winds as a result of the Coriolis effect
    C. at an angle to prevailing winds as a result of the Coriolis effect
    D. at an angle to prevailing winds as a result of the Doppler effect

23. A sargassum is _____.
    A. a flat-topped seamount
    B. the tendency of current directions to be more greatly deflected away from wind direction with depth because of the Coriolis effect and multiple layers of surficial shear within the water
    C. an isolated pocket of water with swirling currents
    D. a tropical alga (seaweed) that accumulates in dense clumps in a region of the Atlantic where currents are still

24. Ekman's spiral is _____.
    A. a flat-topped seamount
    B. the tendency of current directions to be more greatly deflected away from wind direction with depth because of the Coriolis effect and multiple layers of surficial shear within the water
    C. an isolated pocket of water with swirling currents
    D. a tropical alga (seaweed) that accumulates in dense clumps in a region of the Atlantic where currents are still

25. A guyot is _____.
    A. a flat-topped seamount
    B. the tendency of current directions to be more greatly deflected away from wind direction with depth because of the Coriolis effect and multiple layers of surficial shear within the water
    C. an isolated pocket of water with swirling currents
    D. a tropical alga (seaweed) that accumulates in dense clumps in a region of the Atlantic where currents are still

26. An eddy is _____.
   A. a flat-topped seamount
   B. the tendency of current directions to be more greatly deflected away from wind direction with depth because of the Coriolis effect and multiple layers of surficial shear within the water
   C. an isolated pocket of water with swirling currents
   D. a tropical alga (seaweed) that accumulates in dense clumps in a region of the Atlantic where currents are still

27. If currents are largely directed toward shore, the area is likely to experience _____.
   A. upwelling   B. downwelling

28. The most prominent force inducing tides on the Earth is the _____.
   A. Sun's gravitational pull
   B. Moon's gravitational pull
   C. electromagnetic attraction between the Earth and Sun
   D. gravitational attraction between ocean waters and the continents on Earth

29. The magnitude of tidal effect (lateral extent of the intertidal zone between low and high tides) is affected by _____.
   A. the slope of the coastline
   B. the position of the Sun
   C. Both of the above.
   D. None of the above.

30. Tidal reach is affected by _____.
   A. the slope of the coastline
   B. the position of the Sun
   C. Both of the above.
   D. None of the above.

31. The gravitational pull of the Moon on Earth's sublunar tidal bulge causes Earth's _____.
   A. orbital rate around the Sun to increase
   B. orbital rate around the Sun to decrease
   C. rotation rate to increase
   D. rotation rate to decrease

32. A wave will disturb the water to a depth equal to _____.
   A. its wavelength
   B. its surface amplitude
   C. the bottom of the sea floor
   D. one-half of its wavelength

33. Ripple marks in sediment caused by ocean waves are found _____.
    A. at all depths
    B. only within the tidal range
    C. only above the wave base
    D. only below the wave base

34. Wave crests _____.
    A. refract (bend) so as to impact the shore in a nearly parallel manner
    B. refract so as to impact the shore nearly perpendicular to the shoreline
    C. reflect directly backward from the angle at which they approached the shoreline
    D. are reflected off the shoreline, but never refract

35. Rip currents flow_____.
    A. directly toward the shoreline
    B. directly away from the shoreline
    C. parallel to the shoreline

36. Longshore currents flow _____.
    A. directly toward the shoreline
    B. directly away from the shoreline
    C. parallel to the shoreline

37. A swimmer caught in a current heading out to sea should swim _____.
    A. in direct opposition to the current
    B. in the direction of the current
    C. perpendicular to the current (parallel to the shoreline)

38. Sand groins _____.
    A. have solved the problem of beach drift
    B. increase erosion rates in the region upcurrent from the groin
    C. increase erosion rates in the region downcurrent from the groin
    D. increase depositional rates in the region downcurrent from the groin

39. Because of wave refraction, erosion along an irregular coastline is _____.
    A. even along the length of the coast
    B. greatest in headlands (points of land projected out toward the sea)
    C. greatest in bays

40. When sea level rises, the ocean may invade a river valley, producing a nearshore body of water of mixed and variable salinity termed a(n) _____.
    A. atoll
    B. lagoon
    C. estuary
    D. jetty

Chapter 19
A Hidden Reserve: Groundwater

Learning objectives

1. Students should broadly understand the hydrologic cycle driven by evaporation, precipitation, infiltration, and flow. Groundwater flows from regions of recharge to regions of discharge, driven by gravity and water pressure.

2. Pores occur in all geologic materials. If pores are sufficiently connected to form conduits for flow, the material is permeable and can serve as an aquifer (impermeable rocks and sediments are termed aquitards). The water table separates an unsaturated zone (wherein pores are mostly filled with air) from a saturated zone (where water permanently fills the pores) within an unconfined aquifer (one that is not bounded above by an aquitard). Locally, the water table is a subdued mimic of surface topography, rising slightly in hillsides due to increased recharge.

3. Permeability of an aquifer adjusted for the viscosity of the fluid within pores is termed hydraulic conductivity. Darcy's law states that groundwater flux per unit cross-sectional area of an aquifer is equal to the slope of the water table multiplied by hydraulic conductivity.

4. Groundwater can both dissolve and precipitate minerals at or below the surface. Water with excessively high concentrations of dissolved ions is undrinkable. Hard water contains a relatively great concentration of dissolved calcium and magnesium.

5. Pumping water from a well produces a cone of depression in the water table around the well base. Excessive pumping may permanently lower the water table in a region. Use of groundwater at rates that make it a nonrenewable resource is termed groundwater mining.

6. A variety of human products (pesticides, sewage, fertilizer, gasoline, chemicals) have caused serious groundwater contamination problems. Wells can be contaminated by the reversal of groundwater flow in regions that have been overpumped.

7. Caves are etched in limestone by dilute solutions of carbonic acid at or just below the water table. Carbonic acid is introduced into groundwater by dissolved carbon dioxide in rainwater. Water dripping through caves produce a variety of calcitic deposits termed speleothems (including the familiar stalactites and stalagmites). The collapse of caves produces sinkholes; a region dominated by sinkholes is said to exhibit karst topography.

Answers to review questions

1. What factors affect how much water infiltrates into the groundwater in a given region?
    Evaporation rates, the porosity and permeability of the surficial material (rock, sediment, or soil), depth to the water table, vegetation cover, and the saturation state of the pores above the water table (due to recent rains) are all factors that affect infiltration.

2. How does submarine groundwater help keep the Earth temperate?
   Water has great heat capacity. When the Earth is unusually hot, it takes much heat to raise the temperature of groundwater, and groundwater can transport heat away from its source.

3. What types of materials are most porous? Which are least porous?
   Well-sorted, well-rounded sediments have the greatest porosity. Unfractured igneous and metamorphic rocks have the least porosity.

4. Distinguish between primary and secondary porosity.
   Primary porosity is the rock volume (or proportion) made up of pores (which may be filled with air or water) subsequent to deposition (sedimentary rocks or sediments) or formation. Secondary porosity consists of new void space introduced by dissolution or fracturing.

5. What three factors affect the permeability of a substance?
   The number, size, and straightness of conduit pathways through a substance all (positively) affect permeability.

6. How do porosity and permeability differ? Give examples of substances with high porosity but low permeability.
   Porosity describes the proportion of rock or sediment volume comprised of pores, whereas permeability is a measure of the connectedness of pores and the ability of these pores (and fractures) to form conduits through the material. Fine-grained rock, such as mudstone, may have abundant pore space, but the pores are tiny and poorly connected, so it is difficult for a liquid to flow through mudstone, which is considered an aquitard (rock of very low permeability).

7. What factors affect the level of the water table?
   Humidity and great amounts of rainfall cause the water table to rise relatively close to the surface. Locally, the water table will lie closer to the surface in topographic valleys than on hills.

8. What factors affect the flow direction of the water below the water table?
   Groundwater flows from regions of recharge (infiltration) to those of discharge (expulsion into a stream valley or lake), driven by gravity and water pressure.

9. How does the rate of groundwater flow compare with that of moving ocean water or river currents?
   Groundwater flow is very slow compared to currents in surface water, generally moving at less then one and one-half meters per day.

10. How do hydrogeologists measure the rate of water flow?
    Rates of flow can be measured by injecting a dye or radioactive tracer into the groundwater and observing how long this marked water takes to travel a known distance or by observing the proportion of remaining radioactive carbon-14 and comparing this with the distance the water has traveled since recharge.

11. What does Darcy's law tell us about how the gradient and permeability affect discharge?

The discharge rate through a specified cross-sectional area is equal to this area multiplied by the product of head gradient and hydraulic conductivity (permeability adjusted for fluid viscosity).

12. Why is "hard water" hard?

Hard water has great amounts of dissolved calcium and magnesium, which may form precipitate minerals.

13. How does excessive pumping affect the local water table?

Pumping will depress the water table around the pump, forming a cone of depression.

14. How is an artesian well different from an ordinary well?

An artesian well penetrates into a confined aquifer and produces upward flow (which may reach the surface) without any pumping.

15. Explain what makes a geyser erupt.

Water under pressure within fractures in hot rock becomes superheated (above boiling temperature at surface pressure). Eventually, some water will be heated enough to form steam, which will rise up through a conduit to the surface, pushing out some water in its path. This greatly reduces pressure below, so that an abundance of water will quickly expand into steam, explosively ejecting a great quantity of water as it rises to the surface.

16. Is groundwater a renewable or nonrenewable resource? Explain how the difference in time frame changes this answer.

The abundance of water and rainfall on this planet, combined with the slow but persistent travel of groundwater, allows groundwater to be considered a renewable resource on long time scales of many thousands of years. However, locally, on shorter time scales, overuse and pollution of groundwater have led to the consumption and destruction of usable groundwater at rates that are too fast to be ameliorated by natural recharge.

17. Describe some of the ways in which human interference can adversely affect the water table.

Overpumping and the diversion of surface water from areas of recharge have caused the water table to be locally depressed.

18. What are some sources of groundwater contamination? How can it be prevented?

One natural source of contamination (from a human perspective) arises from the rocks through which groundwater flows, through the introduction of unfavorable dissolved ions or gases. Human contamination includes agricultural and industrial chemicals, petroleum, sewage, radioactive waste, and acids from mine drainage. All such wastes should be stored within impermeable bedrock or clay-rich soil to prevent flow from the site.

19. What four factors are typical of regions where caves form?
Caves tend to form in areas where there is a thick body of limestone above sea level with substantial rain and warm temperatures.

Test bank

1. Sinkholes are a concern for residents whose dwellings are constructed atop _____.
   A. sandstone B. shale
   C. limestone D. granite

2. The majority of fresh liquid water within the Earth exists in _____.
   A. lakes B. rivers and streams
   C. pores within rock and sediment D. atmospheric clouds

3. No solid Earth materials consist of greater than 50% porosity.
   A. true B. false

4. Unconsolidated sediment typically has greater porosity than lithified rock which forms from it.
   A. true B. false

5. Primary porosity may be reduced by _____.
   A. compaction of grains
   B. cementation of grains
   C. Both of the above processes reduce primary porosity
   D. None of the above.

6. Groundwater _____.
   A. does not affect the porosity of the rock and sediment through which it flows
   B. uniformly increases porosity due to dissolution of grains
   C. uniformly decreases porosity due to deposition of minerals into pores
   D. may increase or decrease porosity through dissolution or deposition

7. All water within large pores is considered mobile.
   A. true B. false

8. Material through which water readily flows is termed _____.
   A. fluent B. porous C. permeable

9. If a material is porous, it _____.
   A. will be permeable as well
   B. will be impermeable
   C. may be permeable or impermeable

10. If a material is highly permeable, it _____.
    A. must be porous
    B. cannot be porous
    C. must be fractured
    D. must either have substantial porosity or be fractured

11. In unfractured rock and sediment, water molecules usually take a _____.
    A. straight path   B. wandering path   C. circular path

12. An impermeable layer of rock or sediment is termed a(n) _____ in hydrogeologic contexts.
    A. aquitard   B. confined aquifer
    C. unconfined aquifer   D. unsaturated zone

13. Rocks or sediments with water-filled pores, wherein upward percolation is blocked by an overlying, impermeable layer, is termed a(n) _____.
    A. aquitard   B. confined aquifer
    C. unconfined aquifer   D. unsaturated zone

14. Rock or sediment between the water table and the land surface represents a(n) _____.
    A. aquitard   B. confined aquifer
    C. unconfined aquifer   D. unsaturated zone

15. Rock or sediment that bears water, which has the water table as its upper limit, is termed a(n) _____.
    A. aquitard   B. confined aquifer
    C. unconfined aquifer   D. unsaturated zone

16. The topography of the water table _____.
    A. is unaffected by local surface topography
    B. precisely mimics the topography of the ground surface
    C. is a subdued (less steeply sloping) mimic of surface topography
    D. is an exaggerated (more steeply sloping) mimic of surface topography

17. Unless it has recently rained, there is no water within pores in the unsaturated zone.
    A. true   B. false

18. Perched water tables occur _____.
    A. above the regional water table, within permeable rock or sediment
    B. above the regional water table, within impermeable rock or sediment
    C. below the regional water table, within impermeable rock or sediment
    D. below the regional water table, within permeable rock or sediment

19. Which statement about recharge areas is incorrect?
    A. They typically are elevated with respect to neighboring areas.
    B. They are regions of relatively high precipitation.
    C. They are synonymous with "discharge areas."
    D. They are areas where water infiltrates the sediment from above.

20. Freshwater lakes are always discharge areas.
    A. true   B. false

21. The rate of groundwater flux per unit area through a body of rock or sediment depends _____.
    A. only upon the slope of the water table locally
    B. only upon the porosity of the rock or sediment
    C. upon the slope of the water table and the porosity of the rock or sediment
    D. upon the slope of the water table and the permeability of the rock or sediment

22. The relationship governing the rate of groundwater flux was discovered by _____.
    A. Alfred Wegener   B. Isaac Newton
    C. Henri Darcy   D. Charles Richter

23. In groundwater, _____.
    A. dissolved ions are rarely, if ever, found
    B. calcite is more soluble than quartz
    C. minerals may be dissolved but never precipitate out of solution
    D. precipitation may take place but only if the mineral in question is undersaturated

24. Hard water results from relatively high concentrations of dissolved _____.
    A. calcium and magnesium   B. francium and cesium
    C. sodium   D. potassium

25. An ordinary (water-producing) well will result whenever the base of the well extends below the water table.
    A. true   B. false

26. The elevation of the water table _____.
    A. is a constant for a given area so long as the topography remains the same
    B. may rise during times of drought and sink during rainy periods
    C. may rise during rainy periods and sink during droughts

27. A dry well will always result whenever the base of the well is above the water table.
    A. true   B. false

28. Pumping vast quantities of water locally _____.
    A. raises the local water table
    B. lowers the local water table, forming a cone-shaped depression
    C. lowers the local water table, forming a cylindrical depression
    D. does not affect the water table

29. All artesian wells _____.
    A. have a base lying within an aquitard
    B. produce water that flows to the surface without any pumping
    C. have a base lying within a confined aquifer
    D. have a base lying within an unconfined aquifer

30. Any place where groundwater flows from the surface of the Earth is termed a _____.
    A. flowing artesian well   B. geyser
    C. spring   D. recharge area

31. Water flowing from hot springs _____.
    A. contains more dissolved minerals than does water flowing from cool springs
    B. contains less dissolved minerals than does water emanating from cool springs
    C. is never hotter than 40° C, so it is always safe to bathe in
    D. only occurs in regions of active volcanism

32. A periodic explosive eruption of steam and water from within the ground up through the surface is termed a _____.
    A. flowing artesian well   B. geyser
    C. spring   D. recharge area

33. Most of the freshwater utilized globally is pumped from the ground.
    A. true   B. false

34. Groundwater mining refers to _____.
    A. acid drainage from mines polluting groundwater
    B. legal squabbling over groundwater rights ("It's mine!")
    C. pumping out water faster than it can be replenished by recharge
    D. prospecting an area to find a suitable spot for a well

35. The city of Venice, Italy is submerged underwater because _____.
    A. of the rise of sea level after the melting of glaciers after the last ice age
    B. the city was built underwater as a means of defense against ground troops from neighboring city-states
    C. the land it is built on submerged after too much water was pumped from the ground
    D. of global warming melting polar ice fifty years ago

36. If groundwater is clear, it can be assumed that it is drinkable.
    A. true   B. false

37. Caves are carved by groundwaters that contain a solution of _____.
    A. hydrochloric acid B. hydrofluoric acid
    C. sulfuric acid D. carbonic acid

38. Most dissolution of bedrock to form caves takes place _____.
    A. above the water table
    B. just below the water table
    C. greater than 10 m below the water table

39. Caves are produced within _____.
    A. all bedrock lithologies B. limestone
    C. sandstone D. shale

40. Topography dominated by depressions formed by the collapse of caves is termed _____.
    A. valley-and-ridge B. karst C. horst-and-graben

Chapter 20
An Envelope of Gas: Earth's Atmosphere and Climate

Learning objectives

1. Students should know the present and past compositions of Earth's atmosphere. A short-lived initial atmosphere arose during formation of the Earth and would have been dominated by the most common gases within the solar system as a whole, hydrogen and helium. However, Earth was not massive enough to retain these light gases for long. Volcanic outgassing produced the next atmosphere, dominated by water vapor, along with carbon dioxide and sulfur dioxide, and lesser amounts of nitrogen with trace amounts of other gases. Earth soon cooled, and the water vapor from outgassing condensed, forming the oceans. Most of the sulfur dioxide and carbon dioxide went into solution within the oceans, but nitrogen remained. Biogenic photosynthesis gradually added oxygen to the ocean, which later effervesced into the atmosphere, resulting in the nitrogen- and oxygen-dominated mixture that presides today.

2. The atmosphere is divided into four layers from bottom to top, distinguished by changes in direction of the thermal gradient of the atmosphere with increasing altitude. These layers are separated by relatively thin pauses, in which temperature is constant with altitude. The lowest layer, the troposphere, features the phenomena that we call weather, driven by solar-derived heat reradiated at the surface and convected upward. The warmest air is at the base of the troposphere; temperature decreases with altitude. After a pause of constant temperature, temperature increases through the stratosphere, in which oxygen molecules are split and recombined to form ozone, which is heated through its ability to absorb ultraviolet radiation from the Sun. After a second pause of constant temperature, temperatures once again decrease with altitude in the mesosphere, then are constant within a third pause, then increase again within the thermosphere due to absorption of solar radiation.

3. Within the troposphere, global convection is driven by hot air rising at the equator and cold air sinking at the poles, giving rise to three rows of convection cells in either hemisphere (tropical Hadley, temperate Ferrel, and polar). For example, warm moist air rises at the tropics, and cools and sheds rain as it heads in either direction. At 30° latitude, the cold air (now dry) sinks and heats up, forming high pressure and dry conditions in the horse latitudes. The tropical (Hadley) cell is completed by the return of air to the equator along the surface.

4. The Coriolis effect modifies the convection cell flow; thus surface winds in the northern tropics are primarily from the northeast, rather than due north. The Coriolis effect also modifies the flow of air at the top of the troposphere, above the convection cells, so that the flow of air at high altitude toward the poles is deflected east, forming the jet stream. The Coriolis effect additionally produces cyclones about low-pressure regions and anticyclones about high-pressure regions.

5. Weather results from the interactions of air masses. Adjacent air masses may differ substantially in temperature, pressure, and moisture content. Boundaries between air masses are termed fronts.

6. Clouds form as tiny droplets of condensation about aerosol nuclei. Most commonly they are produced when an air mass rises, adiabatically expands, and cools. If water droplets within clouds become too massive to resist gravity, they fall as rain.

7. Climate is a function of latitude, altitude, proximity to standing water, presence and character of nearby ocean currents, topography, and proximity to zones of convergence or divergence (southern or northern extremes of tropospheric convection cells).

Summary from the text

Soon after Earth formed, its atmosphere consisted mostly of hydrogen and helium. This was eventually replaced by an atmosphere containing high concentrations of water vapor, carbon dioxide, and sulfur dioxide, gases erupted from volcanoes.

After the oceans formed, much of the carbon dioxide dissolved in the ocean and then became trapped in limestone. When photosynthetic organisms evolved, they produced oxygen, and the concentration of this gas gradually increased. Our atmosphere today consists of volcanic gas modified by interactions among the components of the Earth system.

Terrestrial life on Earth could not evolve until after the atmosphere contained enough ozone to protect the surface from harmful ultraviolet radiation.

Air, the mixture of gases comprising the atmosphere, consists mostly of nitrogen (78%) and oxygen (21%). Several other gases, including carbon dioxide and methane, occur in trace amounts. These are greenhouse gases, whose concentration affects the temperature of the atmosphere. The atmosphere also contains aerosols, tiny solid particles or liquid drops.

Air pressure decreases with elevation. Thus, 90% of the air in the atmosphere occurs below an elevation of 16 km.

When air pressure decreases (which can occur when air rises), the air expands and cools — a process called adiabatic cooling. If air is compressed, it heats up — a process called adiabatic heating.

Air generally contains water. The ratio between the measured water content and the maximum possible amount of water that the air can hold at a given temperature is its relative humidity (expressed as a percentage).

The atmosphere is divided into layers, separated from each other by pauses. In the lowest layer, the troposphere, temperature decreases with elevation. The troposphere convects; its air movement causes weather. In the next layer, the stratosphere, temperature increases with elevation, and air does not convect. The stratosphere is overlain by the mesosphere and the thermosphere.

Air circulates on two scales, global and local. Winds blow because of pressure gradients: air moves from regions of higher pressure to regions of lower pressure. The energy that drives atmospheric circulation ultimately comes from solar radiation.

High latitudes receive less solar energy, or insolation, than low latitudes. This contrast in temperature initiates convection in the atmosphere; warm air rises at the equator, and cold air sinks at the poles. Because of the Coriolis effect, air moving north from the equator to the north pole deflects to the east, and in each hemisphere, three convection cells develop (the Hadley, Ferrel, and polar cells). Divergence or convergence zones lie between the cells. The boundary between the Ferrel and polar cells is the polar front.

Where air rises, zones of low pressure develop, and where air sinks, zones of high pressure develop. These zones strongly influence global climate.

Prevailing surface winds — the northeast trade winds, the surface westerlies, and the polar easterlies — develop because of circulation in global convection cells. At zones of divergence or convergence, prevailing winds are weak.

Air pressure at a given latitude decreases from the equator to the poles, causing a poleward flow of air. In the Northern Hemisphere, the Coriolis effect deflects this flow to generate high-altitude westerlies. Over the polar front and the horse latitudes, the pressure gradient is particularly steep, so the high-altitude winds are particularly strong; these strong winds are known as jet streams.

Weather refers to the temperature, air pressure, wind speed, and relative humidity at a given location and time. Much of the weather that we enjoy (or endure) reflects the interaction of air masses, bodies of air with recognizable physical characteristics. The boundary between two air masses is a front. At a cold front, a cold air mass pushes underneath a warm one, while at a warm front, warm air rises over cooler air. If a cold front uplifts a warm front, we have an occluded front.

Air sinks in high-pressure air masses and rises in low-pressure air masses. Because of the Coriolis effect, the air begins to rotate around the center of the mass as a consequence. In a cyclone, air rises and circulates counterclockwise around a low-pressure mass. In an anticyclone, air sinks and circulates clockwise around a high-pressure mass.

Clouds, which consist of tiny droplets of water or tiny crystals of ice, form when the air is saturated with water and contains condensation nuclei on which water condenses. Clouds commonly develop where air rises and adiabatically cools. Any of four lifting mechanisms (convective, frontal, convergence, and orographic lifting) may cause air to rise.

In warm clouds, droplets collide and coalesce to form larger drops that can fall as rain. In cold clouds, droplets evaporate, providing moisture that attaches to ice crystals until the crystals become big enough to fall as snow; this is known as the Bergeron process.

Cumulus clouds are puffy, stratus clouds are sheet-like, and cirrus clouds are wispy and delicate. Large cumulonimbus (rain-producing) clouds form where air is unstable and grow across elevation boundaries.

Thunderstorms begin when cumulonimbus clouds grow large. Friction between air and water molecules creates negative charges at the base of the cloud and positive charges at the top. Positive charges also develop on the ground. Lightning flashes when a giant spark jumps across the charge separation. When lightning heats the air and makes it expand explosively, we hear thunder.

Tornadoes, rapidly rotating funnel-shaped clouds, develop in violent thunderstorms. In North America, tornadoes tend to move from southwest to northeast, most commonly in a region of the Midwest called "tornado alley." Meteorologists use the Fujita scale to classify tornadoes.

Nor'easters are large storms associated with wave cyclones.

Hurricanes, huge rotating storms, originate over oceans between 15° and 20° latitude, where the water temperature is above 27° C. The hurricane is nourished by warm, moist air, because when this air rises, moisture condenses and releases latent heat of condensation. The strongest winds occur in the eye wall, the rotating cylinder of clouds surrounding the relatively calm eye.

Climate refers to the average weather conditions in a region over a year. Climate is controlled by latitude, altitude, and the proximity of water and ocean currents, orographic barriers, and high- or low-pressure zones. Climate classes can be recognized by the vegetation they support. Monsoonal climates occur where there is a seasonal shift in the wind direction.

Climate may change periodically. For example, El Niño, a temporary shift in climate triggered by shifts in the position of high- and low-pressure centers in the Pacific, makes an appearance every few years.

Answers to review questions

1. Where does the ozone in the atmosphere come from, where does it mostly occur, and why is it important?

At an elevation of 30 km above the surface (the ozone layer), the sun's energy can break the bonds of molecular oxygen to form atomic oxygen. Atomic oxygen then combines with molecular oxygen to form ozone, O_3. Ozone absorbs ultraviolet radiation which would be harmful to terrestrial life.

2. Describe the composition of air (considering both its gases and its aerosols). Why are trace gases important?

The gas component of air is predominantly nitrogen and oxygen, with trace amounts of argon, water vapor, carbon dioxide, sulfur dioxide, methane, radon, and numerous others. Aerosols include droplets of water, sulfuric acid, and nitric acid, as well as microscopic clay, pollen, soot from fires, and volcanic ash. Some trace gases in the atmosphere are greenhouse gases, trapping infrared radiation and keeping it from radiating into space, thus helping to warm the Earth.

3. How does air pressure change with elevation? Does the density of the atmosphere also change with elevation? Explain why or why not.

Air pressure and density both decrease with increasing elevation. Density (whether conceived as mass density or particle density) decreases; gas molecules become more scarce in the upper reaches of the atmosphere. Air pressure results from the pushing force that gas molecules exert on one another. The more tightly packed the molecules, the more collisions they experience, equating to greater pressure.

4. Describe the atmosphere's structure from base to top. What characteristics define the boundaries between layers?

The lowest 9–12 km of the atmosphere are the troposphere, dominated by heat radiated out of the Earth at its surface, which warms low-lying surficial

air. This warm air expands and thus rises, and cooler air sinks to take its place. This tropospheric convection drives the atmospheric phenomena that we call weather. Although hot air rises, the base of the troposphere (mean temperature of 18° C) remains considerably warmer than the top (-55° C), just as the surface of the Earth stays warmer than the deep interior (from which heat is rising, in part due to convection in the mantle). Above the troposphere is the stratosphere, a zone in which, after a 10-km span of constant temperature with increasing altitude (the tropopause), temperature increases with altitude. At an elevation of about 47 km, the stratosphere gives way to another pause of static temperature (the stratopause, a thin boundary layer). From 47 to 82 km, the mesosphere reverts the trend of the stratosphere; temperature decreases with altitude, reaching a minimum of -85° C at the top. After the mesopause (upper bound of the mesosphere) temperature increases with altitude in the thermosphere, where gases absorb the incident energy of sunlight traveling on short-wave frequencies. In the thermosphere, gas molecules are so sparse that the atmosphere is not well mixed; therefore, low-density gases, such as hydrogen, are stratified above high-density nitrogen and oxygen.
    The pauses represent local extrema in the temperature profile, points or intervals where temperature is constant with increasing altitude.

5. What is the relative humidity of the atmosphere? What is the latent heat of condensation, and what is its relevance to a thunderstorm or tornado?
    The relative humidity of the atmosphere is its moisture content as a percentage of the maximum possible content (a function of temperature). Latent heat of condensation is the heat released when water droplets form from water vapor. Storms can thus derive heat energy from the formation of rain.

6. Explain the relationship between the wind and variations in air pressure.
    Wind consists of the expansion of air and the movement of air molecules from high-pressure to low-pressure regions.

7. Why do changes in atmospheric temperature depend on latitude and the seasons? Why does global circulation break into three distinct convection belts in each hemisphere, separated by high-pressure or low-pressure belts?
    Earth's rotational axis is tilted, so while the Northern Hemisphere experiences summer (being tilted toward the sun's radiation), the Southern Hemisphere is tilted away from the Sun and experiences winter. The development of seasons due to tilt (and the progression of Earth through its orbit, so that seasons change in a yearly cycle) is not so strong as to produce much effect on the temperature in equatorial regions, which are bathed in warm sunshine year round, but is great enough to produce marked effects in temperate and polar regions. The angle of incidence of solar radiation is much more oblique at the poles than at the equator (the poles never face the Sun directly), so average surface and air temperatures decrease from the equator towards either pole.
    Due to bilateral symmetry across the equatorial plane, it is sufficient to address the second question with respect to the Northern Hemisphere. At the equator, warm air rises, and moves north towards cooler air but is deflected by the Coriolis effect and turns toward the east. At a latitude of 30°, the air mass is

effectively heading due east and has cooled to the point that it sinks. Some of this sunken air completes the tropical (Hadley) cell cycle by regressing to the equator along the surface, deflected toward the west by the Coriolis effect. These returning surficial air currents are the trade winds, which blow from the northeast. The rest of the sunken air heads north at the surface, where it is again deflected by the Coriolis effect towards the east (producing the surficial westerlies familiar to inhabitants of the temperate zone).

Meanwhile, at the pole, cold air hugs the surface and migrates south, deflected by the Coriolis effect towards the west to produce the surficial polar easterlies. At the polar front (about 60° latitude) this cold flow converges with warmer surface air that traveled northward. This collision forces some of the air to rise to the upper reaches of the troposphere, where it divides. One portion heads north (northeast, after Coriolis deflection), sinking at the pole to complete a polar convection cell, and the remainder heads south (which ends up being southwest) to complete a temperate (Ferrel) cell when it sinks at the region of the horse latitudes (30° from the equator).

Rising air leaves a zone of low pressure in its wake, so the equator is a zone of low pressure, as are the polar fronts in each hemisphere. Air sinks at the horse latitudes and at the poles, causing surface pressure to be relatively high.

8. Why do prevailing winds develop at the Earth's surface? Why are there prevailing westerly high-altitude winds, such as the jet stream?

Prevailing surface winds are produced by the convective cells within Earth's atmosphere, moderated by the Coriolis effect. For example, hot air rises at the tropics, and diverges, the Northern Hemisphere portion heads northeast, rather than due north, due to the Coriolis effect. At 30° north, the air is cool enough to sink, and roughly half of this air migrates back toward the equator to complete the tropical (Hadley) cell. En route, the Coriolis effect deflects the southbound surface air toward the west, producing the northeasterly trade winds.

Near the top of the troposphere, pressure is greater at the tropics than at the poles, so tropical air expands poleward. In the course of its journey north in the Northern Hemisphere, the Coriolis effect deflects the air masses towards the east, producing the westerly jet stream.

9. Explain the origin of cyclones and anticyclones, and note their relationship to high-pressure and low-pressure air masses. What is a mid-latitude cyclone?

Regions of low pressure attract air masses toward their center; again the Coriolis effect moderates this flow, sending the air spinning counterclockwise in the Northern Hemisphere about the low-pressure center, creating a cyclone. High-pressure centers radiate air outward, which results, in combination with the Coriolis effect, in clockwise rotating air in the Northern Hemisphere, an anticyclone.

A mid-latitude cyclone, in the northern temperate zone, occurs as a low-pressure air mass traveling from west to east, which develops when air masses shear past one another along a cold front.

10. How does a cold front differ from a warm front and from an occluded front?

 A cold front is a boundary defining the leading edge of a mass of cold air that progresses forward by submerging beneath a warmer air mass. A warm front, conversely, tracks the lateral progress of a warm mass of air over and through a more stagnant cold air mass. Cold fronts travel more rapidly than warm fronts, and occasionally a cold front will catch up to a warm front from behind, separating it from the surface to form an occluded front.

11. Why do clouds form? (Include a discussion of lifting mechanisms.) What are the basic categories of clouds?

 Clouds may form due to an influx of water vapor due to evaporation, an increase of aerosol density that provides nuclei for condensation, or cooling to the point that water vapor becomes oversaturated in the atmosphere. An important source of cooling is adiabatic expansion, which occurs when air masses are lifted. Lifting may occur:
 1. through convection (upward rise of a warm air mass)
 2. at the boundary between air masses of differing temperatures (a front) where, likewise, warmer air will rise over cooler air
 3. at a zone of air flow convergence, where air masses collide head-on, leaving upward and outward the only available path for excess air
 4. at a mountain range, where topography impedes lateral flow of air masses, which must then rise over the top of the range to continue

 A few of the more important variations in cloud geometry include cumulus (puffy), stratus (sheet-like), cirrus (thin and wispy), and cumulonimbus (rain-producing dark clouds, which can grow upward and spread laterally at the tropopause to produce anvil-shaped thunderheads).

12. Under what conditions do thunderstorms develop? What provides the energy that drives clouds to the top of the troposphere? Why does lightning occur in such storms?

 Thunderstorms develop with the growth of cumulonimbus clouds, commonly, in temperate climates, within a body of warm, moist air being influenced by an incoming mass of cool, dry air. Fed by the heat energy of the warm air, the cumulonimbus cloud rises into lower-pressure conditions higher up in the troposphere, where the air expands and experiences adiabatic cooling, which leads to the condensation of moisture. This further growth of the cloud releases latent heat of condensation, allowing the cloud to rise yet higher and experience more cooling and more condensation, until it reaches the tropopause. Here the temperature gradient disappears, and further expansion is directed laterally to form an anvil-shaped storm cloud. Lightning arises because electrical charges build up within the cloud, with positively charged ions in the upper reaches and negatively charged ions at the base of the cloud.

13. What conditions lead to the formulation of a tornado?

 Tornadoes in the Northern Hemisphere form when surface winds from the southeast shear past westerly winds at higher altitude, forming a column of air rotating counterclockwise. Updrafts and downdrafts within the vortex tilt the column and cause it to intersect with the ground surface, where it may skip and bob, or may traverse an extended path.

14. Describe the stages in the development of a hurricane. Describe a hurricane's basic geometry.

A hurricane is a strong cyclone born over warm ocean water. It begins with warm, moist air rising in the tropics, creating low pressure as it rises and being swirled into a vortex due to the Coriolis effect (as in any other low-pressure Northern Hemisphere cyclone). As in an ordinary thunderstorm, rising air brings adiabatic cooling, which brings condensation and the release of latent heat, adding fuel to the building storm. Beyond an arbitrary wind speed threshold, the storm is considered to be a tropical depression. Over warm ocean waters, the strength of a hurricane can continue to build as the ocean provides an abundance of warmth and moisture, which becomes sucked up in the vortex, to produce yet more condensation, and yet more available energy. When winds as strong as 161 km per hour are sustained, the storm is termed a hurricane.

15. What factors control the climate of a region? What special conditions cause monsoons? El Niño?

Latitude, altitude, topography, proximity to the ocean or a large lake, temperature and direction of nearby ocean currents (if there are any), and proximity to regions of consistently high or low pressure all affect climate. Monsoons arise when prevailing winds change direction seasonally, as in India, where northerly winds arising from a Tibetan high-pressure center keep the land dry in the winter, but this high gives way to a low that develops with summer heating, creating southerly winds that allow moist air masses from the Indian Ocean to drench southern Asia.

El Niño results from the displacement of a low-pressure zone, normally found over Indonesia, to the east. This causes surface winds which normally blow from the east of the coast of Peru (causing upwelling of cold, nutrient-rich water which supports commercial fishing industries) to blow westerly, toward the coast. These onshore winds cease coastal upwelling and lead to a variety of unusual weather patterns.

Test bank

1. The very first Earth atmosphere, forming with the Earth itself, was probably dominated by the cosmically abundant gases _____.
    A. nitrogen and oxygen   B. hydrogen and helium
    C. carbon dioxide and methane   D. water vapor and sulfur dioxide

2. Which two gases are most abundantly emitted in volcanic outgassing?
    A. nitrogen and oxygen   B. hydrogen and helium
    C. carbon dioxide and methane   D. water vapor and sulfur dioxide

3. Which two gases are most abundant within Earth's modern atmosphere?
    A. nitrogen and oxygen   B. hydrogen and helium
    C. carbon dioxide and methane   D. water vapor and sulfur dioxide

4. Carbon dioxide is rare within the modern atmosphere because _____.
    A. Earth's volcanoes, unlike those of Venus and Mars, rarely if ever emit carbon dioxide
    B. most of the carbon dioxide from volcanoes was dissolved in the ocean and has been utilized in photosynthesis to produce organic carbon
    C. most of the carbon dioxide from volcanoes was dissolved in the ocean and has been converted to limestone
    D. None of the above; carbon dioxide is the most abundant gas in Earth's atmosphere.

5. Oxygen in Earth's atmosphere built up over time as a result of _____.
    A. volcanic outgassing
    B. photosynthesis
    C. limestone formation in the ocean
    D. solar radiation splitting molecules of ozone in the stratosphere

6. Which two gases are most important in producing the greenhouse effect on Earth?
    A. nitrogen and oxygen
    B. hydrogen and helium
    C. carbon dioxide and methane
    D. water vapor and sulfur dioxide

7. As an air mass rises, it generally experiences _____.
    A. an increase in pressure
    B. a decrease in pressure
    C. no change in pressure

8. As an air mass rises, it generally experiences _____.
    A. an increase in temperature due to adiabatic heating
    B. a decrease in temperature due to adiabatic cooling
    C. an increase in temperature due to increasing proximity to the Sun
    D. no change in temperature

9. A warm air mass may hold more moisture (water vapor) than a cold air mass of the same volume and yet may have lesser relative humidity.
    A. true   B. false

10. To an observer traveling outward from Earth's surface all the way out into space, the temperature of the atmosphere _____.
    A. uniformly decreases due to atmospheric thinning
    B. uniformly increases due to increasing proximity to the Sun
    C. remains relatively constant until the very edge of the atmosphere, where the observer would experience the very cold vacuum that is interplanetary space
    D. decreases, stabilizes, increases, stabilizes, decreases, stabilizes, and, finally, increases until the edge of the atmosphere is reached

11. Atmospheric convection of heat radiated from Earth's surface and the variety of phenomena we call weather occur within which layer of the atmosphere?
	A. mesosphere
	B. stratosphere
	C. thermosphere
	D. troposphere

12. Most atmospheric ozone occurs in which layer of the Earth's atmosphere?
	A. mesosphere
	B. stratosphere
	C. thermosphere
	D. troposphere

13. In which layer of the atmosphere does temperature decrease with altitude because most of its heat is derived from the atmospheric layer immediately below it?
	A. mesosphere
	B. stratosphere
	C. thermosphere
	D. troposphere

14. Which layer of the atmosphere has the lowest density?
	A. mesosphere
	B. stratosphere
	C. thermosphere
	D. troposphere

15. Which layer of the atmosphere is chemically stratified, with the lightest gas molecules riding above heavier molecules?
	A. mesosphere
	B. stratosphere
	C. thermosphere
	D. troposphere

16. Lines connecting points of equal atmospheric pressure on a map are termed _____.
	A. contour lines
	B. isobars
	C. isograds
	D. isotherms

17. Lines connecting points of equal temperature on a map are termed _____.
	A. contour lines
	B. isobars
	C. isograds
	D. isotherms

18. Laterally, at the surface, winds blow from regions of _____.
	A. high temperature to regions of low temperature
	B. low temperature to regions of high temperature
	C. high pressure to regions of low pressure
	D. low pressure to regions of high pressure

19. In global convective circulation, air rises at the _____.
	A. equator and poles
	B. equator and polar fronts
	C. horse latitudes and poles
	D. horse latitudes and polar fronts

20. In global convective circulation, air sinks at the _____.
	A. equator and poles
	B. equator and polar fronts
	C. horse latitudes and poles
	D. horse latitudes and polar fronts

21. Due to the Coriolis effect, air that is spread northward from the equator is deflected _____.
    A. directly back to the equator B. to the east
    C. to the west D. upward to the tropopause

22. Where air sinks, as in the subtropical horse latitudes, the air usually _____.
    A. cools as it sinks
    B. creates high pressure locally
    C. creates low pressure locally
    D. increases in relative humidity as it sinks

23. At zones of divergence and convergence, such as the equator, poles, horse latitudes, and polar fronts, surface winds are generally _____.
    A. easterly
    B. westerly
    C. southerly in the Southern Hemisphere but northerly in the Northern Hemisphere
    D. calm or weak

24. At high altitude within the troposphere, the global pattern of air flow is from the _____.
    A. equator to the poles B. poles to the equator

25. The high-altitude jet stream in the Northern Hemisphere travels from _____ due to the Coriolis effect on northbound air masses.
    A. east to west B. west to east

26. Starting from the equator and moving towards either pole, how many rows of convection cells are present within the troposphere?
    A. one B. two C. three D. four

27. The boundary between two bodies of air with differing characteristics (such as temperature, pressure, or relative humidity) is termed a _____.
    A. contact B. front
    C. unconformity D. Gruber surface

28. As compared to warm fronts, cold fronts move _____.
    A. more rapidly
    B. more slowly
    C. at the same rate but always in opposite directions
    D. at the same rate and always in the same direction

29. Cold fronts and warm fronts tend to deflect one another, so it is impossible for a warm front to become occluded by a cold front.
    A. true B. false

30. Cold air pushes underneath warm air at cold fronts and warm air pushes underneath cold air at warm fronts.
    A. true B. false

31. In the Northern Hemisphere, air travels in a _____ direction around areas of high pressure.
    A. clockwise   B. counterclockwise

32. The development of cyclones and anticyclones about high- and low-pressure regions is a result of the deflection of air currents by _____.
    A. solar wind   B. the stratosphere
    C. the Coriolis effect   C. ion beams originating in the ionosphere

33. Clouds generally form when air _____.
    A. is heated   B. rises
    C. sinks   D. is devoid of aerosols

34. Compared to the low-lying areas which surround them, mountain ranges typically experience _____.
    A. more rainfall
    B. less rainfall
    C. the same amount of rainfall

35. Which cloud type produces rainstorms?
    A. cirrus   B. cumulus
    C. cumulonimbus   D. stratus

36. Which cloud type consists of thin wisps of droplets high in the troposphere?
    A. cirrus   B. cumulus
    C. cumulonimbus   D. stratus

37. Which cloud type has a sheet-like geometry?
    A. cirrus   B. cumulus
    C. cumulonimbus   D. stratus

38. The positive feedback that allows storm clouds to grow upward to the base of the tropopause is due to energy obtained from _____ during cloud formation.
    A. the Coriolis effect   B. the jet stream
    C. latent heat of condensation   C. extra sunlight reflected off the Moon

39. Lightning not only strikes the surface of the Earth, but also travels from cloud to cloud.
    A. true   B. false

40. The strongest winds in a hurricane are found within the central eye.
    A. true   B. false

Chapter 21
Dry Regions: The Geology of Deserts

Learning objectives

1. Students should know that deserts are defined on the basis of aridity; they are characterized by little annual precipitation, lengthy periods of drought, sparse vegetative cover, high rates of evaporation, and an absence of permanent streams of local origin. Deserts can be quite cold, as in the Gobi and the polar regions. Even hot deserts cool off drastically at night, as radiation of heat into space is greater in desert areas than in other environments.

2. Students should know the physiographic factors that produce deserts: descent of cool, dry air in the subtropics, development of rain shadows on the leeward side of mountain ranges, isolation from the ocean within the interior of a massive continent, and cold ocean currents. Many desert areas are so arid because of a combination of these factors.

3. Although chemical weathering rates are slow in arid conditions, chemical dissolution and reprecipitation are responsible for the formation of desert varnish, case hardening, and caliche. Physical weathering and erosion take place through fracturing along joints, wind abrasion, and scouring by flash flood waters during infrequent, but torrential, rains. Wind carries smaller, dust-size particles, abrading rocks to form ventifacts and leaving behind coarser lag deposits of pebbles and boulders. Water from ephemeral streams arising from rainstorms releases its sediment when it reaches a broad plain, producing a wedge of sand and gravel termed an alluvial fan.

4. Many typical desert landforms, such as buttes, cuestas, and mesas, result from cliff retreat, episodic breakdown of the cliff face along vertical joints.

5. The plants and animals of the desert have evolved adaptations to survive in a dry climate, including water storage, deep root systems, and expansive, shallow root systems in plants, as well as nocturnal activity and decreased or concentrated excretion in mammals.

6. In many places of the world, deserts are expanding due to an influx of human populations and agricultural activities, including diversion of water for irrigation and the expansion of grazing into semi-arid regions, which cannot recover fast enough to tolerate herbivorous predation.

Summary from the text

Deserts receive so little rain (less than 25 cm per year) that no more than 15% of their surface is covered by vegetation. Permanent streams cannot survive there. In cold deserts (including polar regions), temperatures stay below 20° C, and in hot deserts, temperatures rise above 35° C.

Desert landscapes do not allow for thick soils, and because they have so little vegetation cover, they are particularly susceptible to erosion by the wind and rain.

Subtropical deserts form between 20° and 30° latitude, rain-shadow deserts are found on the inland side of mountain ranges, coastal deserts are found on the land adjacent to cold ocean currents, continental-interior deserts exist in landlocked regions far from the ocean, and polar deserts form in polar regions (at high latitudes).

Ultimately, plate movements determine whether a section of continental crust lies in a region of dry air.

In deserts, chemical weathering happens slowly, so rock bodies tend to erode primarily by the physical weathering of joints. Nevertheless, chemical weathering does occur, leading to case hardening and the formation of desert varnish. Some soils contain caliche.

Water causes significant erosion in deserts, mostly during heavy downpours. Rain breaks up surface sediment, and large quantities of coarse sediments, including boulders, are carried by ephemeral streams. When the rain stops, these streams dry up, leaving a steep-sided wash (also called an arroyo or wadi).

Wind causes significant erosion in deserts, for it picks up dust and silt and carries them as suspended load and causes sand to saltate (bounce along the ground). Where wind blows away finer sediment, a lag deposit remains, which later settles into a mosaic called desert pavement. Wind-blown sediment abrades the ground, creating a variety of features such as ventifacts and yardangs. Wind erosion can also cause deflation, a lowering of the ground surface, and scour out topographic depressions called blowouts.

Talus aprons form when rock fragments accumulate in an apron at the base of a steep slope. Alluvial fans form at a mountain front where water in ephemeral streams deposits its sediment load. A row of overlapping alluvial fans is called a bajada. When temporary desert lakes dry up, they leave a playa, a flat area covered by a smooth surface of clay and salts.

In some desert landscapes, erosion causes cliff (scarp) retreat of high areas, eventually resulting in the formation of inselbergs ("island mountains"), in some cases surrounded by a pediment. The erosion of stratified rock yields such landforms as buttes, mesas, and cuestas.

Where there is abundant sand, the wind builds it into dunes. Common types include barchan, star, transverse, parabolic, and longitudinal (seif) dunes.

Deserts contain a great variety of plant and animal life. All are adapted to survive extreme temperatures without abundant water.

Changing climates and human practices have resulted in desertification of semi-arid land into deserts. In many parts of the world, desertification has resulted in the loss of valuable farmland.

Answers to review questions

1. What factors determine whether a region can be classified as a desert or not?

Deserts are extremely arid regions marked by sparse vegetation cover (less than 15% of the ground surface), little precipitation (less than 25 cm per year), and high rates of evaporation. Local streams are ephemeral in deserts; only rivers fed by rain and snowfall upstream stay wet throughout the year.

2. Explain why deserts form in (a) the subtropics; (b) the inland side of mountain ranges; (c) coastal regions adjacent to cold ocean currents; (d) the interiors of large continents; (e) polar latitudes.

a. Subtropical deserts persist because of Hadley cell convection. Warm, moist air rises over the tropics and spreads laterally, shedding rainfall as it cools, by the time the air is above the subtropics, its moisture content is very low and it cools enough to sink. As the cool, dry air sinks, it becomes warm, dry air. Warm air has greater capacity to hold water vapor than does cool air, so this hot dry air mass soaks up available moisture, causing the high evaporation rates characterizing subtropical deserts.

b. Rain-shadow deserts form on the leeward side of mountain ranges. As moisture-laden air masses encounter mountains, they are forced to rise over the topography, cooling and shedding moisture as they journey upward on the windward side of the mountain. Most of their moisture spent, these air masses heat up as they descend down the leeward side of the mountains, and their relative humidity decreases even further.

c. Cool-water currents absorb heat from the air above them, keeping the temperature of the air, and its capacity to carry moisture, low.

d. Deserts form in continental interiors because moisture derived from the ocean is likely to be shed on the journey to the interior, leaving little moisture available for land in the center of a large continent.

e. Polar regions are arid because cold air can hold only small amounts of moisture and because air sinks at the poles, as in the subtropics.

3. Have today's deserts always been deserts? Keep in mind the consequences of plate tectonics as you answer this question.

The positions of deserts have shifted through time, driven by plate tectonic movements. Collisions of continental masses produce mountains, which may form rain shadows and can lead to the assembly of large continents with dry interiors. Plate motion may pull a continent into the subtropics, and positions of the continents can influence current patterns.

4. How do weathering processes in deserts differ from those in temperate or humid climates?

In desert conditions, chemical weathering takes place at slow rates, so physical weathering processes provide most of the loose material which may be eroded away.

5. Describe how water modifies the landscape of a desert. Be sure to discuss both erosional and depositional landforms that involve water.

Water from rare, but occasionally intense, rainfall is an important agent of physical weathering and erosion in deserts. Without much vegetation to hold sediment in place, it can be dislodged merely by the impact of rain. The ground can quickly become saturated with water, and a sheetwash may course across the land surface, carrying loose sediment with it. The water heads for arroyos, and a flash flood of water and entrained sand, mud, and gravel travels downstream at high velocity. The sediment load scours the channel as it travels, and grains abrade one another to produce finer particle sizes which can be carried farther by the ephemeral stream. When the stream channel terminates in a planar region, it breaks into distributary channels and deposits sand and gravel in a wedge-shaped deposit termed an alluvial fan.

6. Explain the ways in which wind transports sediment in a desert.
   Wind can support fine dust as a suspended load, which may be blown outside the desert itself to form a loess deposit. Sand is transported as a bed load through the rolling and saltation of grains.

7. Explain how the following features form: (a) desert varnish; (b) desert pavement; (c) ventifacts; (d) yardangs; (e) blowouts.
   a. Desert varnish forms when water dissolves iron and manganese from the interior of a rock and pulls them to the surface, where the water evaporates, leaving behind a rind of iron oxides and manganese oxides.
   b. Desert pavements form from the coarse pebbles, cobbles, and boulders in lag deposits which remain after sand and finer material have been blown away from an area. These coarse grains will pack together to form a mosaic, or pavement, at the surface.
   c. Ventifacts are rocks with smooth surfaces, produced through wind abrasion, that meet along sharp edges. The multiple facets form because either the rock has shifted its orientation with respect to prevailing winds, or the direction of the prevailing winds has changed during its history.
   d. Yardangs form when rock that is resistant to the wind's abrasion overlies less resistant rock, leaving an extended cap of resistant rock supported by a thin pedestal of the less resistant material.
   e. Blowouts are depressions formed by wind scour.

8. Describe the appearance and process of formation of the different types of depositional landforms that develop in deserts.
   Material that breaks off of a hillside tumbles down to form a talus apron of angular grains ranging from pebbles to large boulders at the base of the hill.
   Alluvial fans are wedge-shaped bodies of relatively coarse sediment that arise at the mouths of channels that scour into mountains. Upon reaching flat land, the channel breaks into distributaries, which, after being fed by upstream flooding, deposit lenses of sand and gravel.
   Playas are flat-lying deposits of clay and evaporite minerals that form when a temporary lake evaporates.
   Winds may settle to produce loess deposits and can push sand into elongate, star-shaped, or arcuate piles termed dunes.

9. Describe the process of cliff (scarp) retreat and the landforms that result from it. What factor determines whether a mesa or a cuesta forms in a region?
   Cliff retreat involves the loss of material along a cliff face due to weathering along vertical joints. The retreat of cliff faces produces isolated uplands, including broad mesas, tall but narrow chimneys, and buttes of intermediate lateral extent; these three landforms have flat tops because they are formed from flat-lying strata. Cuestas are similar to mesas, with the exception that their upper surfaces are gently sloped, because they form from strata that dip away from horizontal.

10. How do sand dunes develop? What are the various types of dunes? What factors determine which type of dune develops in a given location?

Sand dunes are piles of sand that are produced and modified by the wind; they typically have an asymmetric profile, with saltating grains traversing across the more gently sloped windward side of the dune and accumulating on the steeper lee side. Various shapes of dunes have been given the names barchan, star dune, transverse dune, and longitudinal dune (or seif). Sand dune geometry is controlled by wind speed, consistency of wind direction, and sediment supply.

11. Discuss various adaptations that life forms have evolved in order to be able to survive in desert climates.

Plant adaptations to desert climate are numerous and varied. Some plants produce seeds with thick coats that lay dormant until they spring into life and reproduce during a downpour. Root systems may extend deep within the ground to reach the water table or may shoot out laterally just beneath the surface to soak up as much rain as possible during downpours. Succulents and cacti store water internally in thick, fleshy conductive tissue surrounded by an unusually thick, water-resistant waxy cuticle. Needles on cacti protect water stores from potentially marauding animals.

Frogs may lie dormant in a burrow until heavy rains unleash a flurry of amphibian reproductive activity. Most mammals and toads avoid the heat and high rates of evaporation present during the day by being nocturnal. Lizards and snakes seek the shade of rocks during midday. Birds find shade in bushes and in hollows within cacti. Some mammals do not sweat, keeping evaporative water loss to a minimum; others produce highly concentrated urine. Large ears are common in desert mammals, serving to radiate excess body heat (which is conveyed to the ears by blood vessels).

12. What is the process of desertification, and what causes it?

Desertification is the expansion of deserts into land that was previously nondesert. This process is hastened by agricultural grazing and plowing of the land, which remove vegetation, as well as by natural droughts and the diversion of water to provide agricultural irrigation (or to meet the needs of a rapid influx of population).

Test bank

1. To qualify as a desert, a region must be _____.
    A. hot, with a mean annual temperature greater than 25° C (77° F)
    B. arid, with less than 25 cm annual precipitation and very low relative humidity
    C. both hot and arid
    D. either hot or arid

2. Vegetation in the desert consists of _____.
    A. cacti only
    B. extensive grasslands
    C. scattered, scrub-like plants with adaptations to obtain and retain as much water as they can
    D. forests of tall, broadleaf trees

3. The highest recorded temperature on Earth was in a _____.
   A. tropical rainforest in Brazil
   B. low latitude, high elevation desert in Mexico
   C. high latitude, high elevation desert in Mongolia
   D. low latitude, low elevation desert in Libya

4. During heavy rainstorms, rates of physical weathering and erosion are _____.
   A. greater in humid climates than in deserts, because dry desert soils can soak up all of the available moisture
   B. greater in deserts than in humid climates, because vegetation tends to hold soil together
   C. greater in humid climates than in deserts, because vegetation tends to break up the soil
   D. about equally fast in deserts and in humid climates

5. Most hot deserts _____.
   A. retain their high temperatures throughout the night, because there is no vegetation to absorb heat from the sediment at the surface
   B. cool off greatly at night, because of sparse vegetation and little cloud cover
   C. cool off slightly at night, by no more than 10° C

6. The most expansive hot deserts on Earth occur _____.
   A. in the tropics, at or near the equator
   B. in the subtropics, between 20 and 30° north or south of the equator
   C. in the temperate zone, between 30 and 50° north or south of the equator

7. At 30° north latitude, at the northern edge of tropical (Hadley) cell convection in the Northern Hemisphere, _____.
   A. cool, dry air sinks, becoming drier as it heats up
   B. warm, moist air rises, increasing in relative humidity as it rises
   C. warm, dry air rises, becoming cooler
   D. cool, moist air sinks, providing abundant rainfall

8. Evaporite deposits in coastal, subtropical regions derived from the evaporation of seawater stranded above high tide are termed _____.
   A. mirages   B. caliches
   C. sabkhas   D. yardangs

9. Desert climate associated with a rain shadow is found _____.
   A. on the windward side of mountain ranges
   B. on the leeward side of mountain ranges
   C. in the middle of flat plains
   D. along continental coastlines

10. Deserts in coastal regions are most likely to be found where the currents are _____.
    A. warm
    B. cold
    C. headed toward the east
    D. headed toward the west

11. Some of the physiographic features that cause extreme aridity may be found together within a single desert, but the deserts found in the western United States are primarily due to being located _____.
    A. in the subtropics
    B. within the rain shadow of a mountain range
    C. near the center of a large continent
    D. next to a cold ocean current

12. The Sahara of Africa is a desert primarily because it is located _____.
    A. in the subtropics
    B. within the rain shadow of a mountain range
    C. near the center of a large continent
    D. next to a cold ocean current

13. The Atacama of Chile is a desert primarily because it is located _____.
    A. in the subtropics
    B. within the rain shadow of a mountain range
    C. near the center of a large continent
    D. next to a cold ocean current

14. The Gobi of Mongolia is a desert primarily because it is located _____.
    A. in the subtropics
    B. within the rain shadow of a mountain range
    C. near the center of a large continent
    D. next to a cold ocean current

15. Most regions that are now deserts _____.
    A. have been deserts throughout the geologic past
    B. have experienced other climates in the geologic past

16. As compared to humid climates, rates of chemical weathering in deserts are _____.
    A. faster
    B. slower
    C. very similar

17. The iron oxides and manganese oxides that produce desert varnish on the exterior surfaces of rocks are derived from ions _____.
    A. in the interiors of the rocks, which are transported outward by the capillary action of water
    B. in the interiors of the rocks, which are transported outward by microorganisms
    C. that occur naturally in rain water and precipitate out as oxides after the rain evaporates
    D. that occur naturally on the surfaces of carbonate and silicate rocks

18. Case hardening _____.
    A. is the process that produces desert varnish
    B. bears no similarity to the process that forms desert varnish
    C. is similar to desert varnish in that it produces iron oxides and magnesium oxides at the surface
    D. forms in a manner similar to desert varnish but produces a calcite or quartz crust at the surface rather than an oxide stain

19. Native Americans produced petroglyphs by etching into _____.
    A. rocks that had been subjected to case hardening
    B. rocks that had been coated with desert varnish
    C. dark basalts
    D. obsidian

20. Caliche forms in a desert climate through the dissolution and reprecipitation of _____ during and after rainstorms.
    A. manganese oxide   B. quartz
    C. calcite   D. iron oxide

21. In most deserts, the scarcity of rainfall means that most of the physical weathering and erosion is accomplished by the wind.
    A. true   B. false

22. An arroyo is a steep-sided valley produced by _____.
    A. normal faulting
    B. wind erosion
    C. scouring erosion by water and sediment during flash floods
    D. cliff retreat

23. Saltation of sand involves _____.
    A. spherical grains rolling along the surface of dunes
    B. grains hopping into the air, traveling for a short distance, and returning to the ground
    C. grains being carried long distances (up to tens of kilometers) by strong gusts of wind
    D. the mixture of quartz sand with salt, which forms from the evaporation of rainwater

24. A rock that has been abraded by windborne particles on multiple faces, forming a sharp edge in between is termed a _____.
    A. blowout   B. ventifact   C. wadi   D. yardang

25. A bowl-shaped depression created by a vortex of scouring wind is termed a _____.
    A. blowout   B. ventifact   C. wadi   D. yardang

26. A mushroom-shaped landform consisting of a column of less resistant rock supporting a broader extent of wind-resistant rock is termed a _____.
    A. blowout   B. ventifact   C. wadi   D. yardang

27. A(n) _____ is a wedge of sand and gravel deposited by distributary channels that arise when an ephemeral stream reaches a plain at the base of a slope.
    A. alluvial fan   B. ventifact   C. wadi   D. yardang

28. A(n) _____ is synonymous with the term "arroyo."
    A. alluvial fan   B. ventifact   C. wadi   D. yardang

29. Geometry of sand dunes is strongly influenced by _____.
    A. the strength of the wind   B. the consistency of wind direction
    C. the abundance of sand   D. All of the above.

30. The distinction between a mesa and a cuesta is _____.
    A. a mesa has a narrow lateral extent, whereas cuestas are more expansive
    B. a mesa is a depositional feature, whereas cuestas are formed primarily by erosion
    C. the upper surface of a mesa is horizontal, whereas the upper surface of a cuesta is sloped
    D. the upper surface of a cuesta is horizontal, whereas the upper surface of a mesa is sloped

Chapter 22
Amazing Ice: Glaciers and Ice Ages

Learning objectives

1. Glaciers are bodies of recrystallized ice that flow in response to gravity. Mountain glaciers travel down slopes; continental glaciers flow because of pressure applied by the weight of ice at their source areas.

2. Mountain and continental glaciers produce a variety of distinctive erosional and depositional features. Students should be familiar with the appearance and mode of formation for some of these, including cirques, arêtes, horns, glacial valleys, ground moraines, end moraines, drumlins, eskers, and kettles.

3. Glacial drift consists of a variety of sediments representing distinct environments. Till is poorly sorted sediment dropped by glaciers when they melt. Meltwater streams produce stratified deposits termed outwash. Meltwater lakes deposit fine sediment in rhythmic couples termed varves. Strong catabatic winds blow fine silt and clay over long distances; these sediments settle out as the wind dies down to form loess.

4. Glaciers have advanced over the continents in the past during broad spans of Earth history termed icehouse conditions, characterized by slow spreading rates along mid-ocean ridges, low atmospheric carbon dioxide content, low sea level, and large continental masses positioned near the poles. Icehouse conditions historically occurred in the early Proterozoic, late Proterozoic, early Paleozoic, late Paleozoic, and late Cenozoic.

5. Within times of icehouse climate, glaciers advance over the continents due to Milankovitch orbital parameters (changes in orbital eccentricity and axial tilt), augmented by an increase in global albedo resultant from widespread ice, interruption of global currents, and the development of coal swamps (which bury organic carbon so it cannot reach the atmosphere as carbon dioxide).

6. The most recent series of glacial advances began about 3 million years ago. Four major episodes of glacial advance and retreat were first recognized in North America from sedimentary evidence, but many more events are recorded by marine sediments (the continental record is much more incomplete due to erosion). Times of minimal glacial extent within broader icehouse climate are termed interglacials; we are currently living in an interglacial that began about 11,000 years ago.

Summary from the text

Ice has a high albedo (reflectivity) and is less dense than liquid water. It is an aggregate of crystals and thus can be considered a rock.

Glaciers are streams or sheets of recrystallized ice that survive for the entire year. Mountain (alpine) glaciers exist in very high regions, and continental glaciers (ice sheets) spread over substantial areas of the continents.

Mountain glaciers include cirques, bowl-shaped depressions on the flank of a mountain, and valley glaciers. Some may form mountain ice caps that cover peaks, and some may spread out into lobes called piedmont glaciers, on the land next to the mountains.

Glaciers form when snow accumulates over a long period of time. With progressive burial, the snow first turns to firn, and then to ice.

Wet-bottom glaciers move by basal sliding over water or wet sediment. Dry-bottom glaciers move by internal flow; ice crystals become plastically deformed, then slide past one another. In general, glaciers move tens of meters per year, but during a surge, they may flow up to 110 m per day.

The upper 60 m of a glacier is too brittle to flow, so it deforms by cracking. Large cracks are called crevasses.

Valley glaciers move because gravity pulls them down slopes, while continental glaciers move because of gravitational spreading (the weight of the ice makes them move out laterally, like syrup flowing over a pancake).

Whether the terminus (toe) of a glacier stays fixed in position, advances farther from the glacier's origin, or retreats back toward the origin depends on the balance between the rate at which snow builds up in the zone of accumulation and the rate at which the glacier melts or sublimates in the zone of ablation. If accumulation exceeds ablation, the glacier advances, while if ablation exceeds accumulation, the glacier retreats.

Icebergs break off glaciers that flow into the sea, and calve off sea ice, which forms when the ocean surface freezes. Sea ice grounded along the coast is called an ice shelf.

Glacial ice can plow over sediment, incorporate sediment, and pluck sediment. The clasts embedded in glacial ice act like a rasp that abrades the substrate.

Mountain glaciers carve numerous landforms, including cirques, arêtes, horns, U-shaped valleys, hanging valleys, and truncated spurs. Glacially carved valleys that fill with water when the sea level rises after an ice age are called fjords.

Glaciers can transport sediment of all sizes (known collectively as glacial drift). Some has been plucked from the substrate, while some falls on the glacier from adjacent mountains. Glacial drift includes till (unsorted sediment), glacial marine (accumulated on the sea floor), glacial outwash (gravels and sand of outwash plains), varves of glacial lake beds, and loess (wind-transported sediments). Lateral moraines accumulate along the sides of valley glaciers, and medial moraines form down the middle of a glacier. End moraines accumulate at a glacier's terminus, and lodgment till beneath a glacier.

Numerous depositional landforms develop in glacial environments or during glacial retreat. These include moraines, knob-and-kettle topography, drumlins, kames, eskers, meltwater lakes, and outwash plains.

The sinking of continental crust as a result of ice loading is glacial subsidence. When the glacier melts away, the sea level rises.

Continental glaciers, as they recede, can dam valleys and cause huge lakes to accumulate. When the ice dams break, a large flood ensues. Such a flood carved the channeled scablands of eastern Washington.

During ice ages, the climate in regions south of the continental glaciers was wetter, and pluvial lakes formed in regions that are now desert. In polar latitudes today, permafrost (permanently frozen ground) exists in periglacial environments.

During the Pleistocene ice age, which began between 3.0 and 2.5 million years ago, large continental glaciers covered much of North America, Europe, and Asia. Huge mammals (now extinct) roamed the land, and modern humans first appeared at this time.

The stratigraphy of Pleistocene glacial deposits preserved on land records five European and four U.S. glaciations, times during which ice sheets advanced, separated from each other by interglacials, times during which ice sheets retreated. The record preserved in marine sediments records twenty to thirty such events.

The geologic record suggests that only four or five ice ages have occurred during Earth history. Long-term causes include plate tectonics and changes in the concentration of carbon dioxide in the atmosphere. Short-term causes include the Milankovitch cycles (caused by periodic changes in Earth's orbit and tilt), a changing albedo, changes in ocean currents, and changes in plankton productivity.

Positive feedback mechanisms drive the climate to become still colder once it has started to cool. Advances and retreats during an ice age are largely regulated by the Milankovitch cycles. Another advance may conceivably happen in the future, though it may be delayed by human-induced global warming.

Answers to review questions

1. What evidence did Louis Agassiz offer to support the idea of an ice age?
  He noticed many European deposits contained boulders, which water could not carry, and were unsorted (water produces sorted deposits because carrying capacity is a function of velocity).

2. How do mountain glaciers and continental glaciers differ in terms of dimensions, thickness, and patterns of movement?
  Continental glaciers are thicker, much more expansive sheets. Mountain glaciers flow down hill as a result of gravity acting on the mass of ice. Continental glaciers move in response to pressure from the weight of material in their thick midsections.

3. Describe the transformation from snow to ice.
  Fluffy snow, when sufficiently buried, packs together and melts in places due to pressure from above. The liquid refreezes to produce more tightly packed firn ($1/4$ air by volume) and ultimately a solid mass of interlocking ice (interrupted by bubbles).

4. Explain how arêtes, cirques, and horns form.
  Cirques are bowl-shaped depressions scoured by mountain glaciers and the sediments they carry. Arêtes are residual, elongate ridges between cirques, and horns are residual, pointed peaks at the intersection of three or more arêtes.

5. Describe the mechanisms that allow glaciers to move, and explain why they move.
  If the bottom surface along which the glacier is traveling is wet, it may slide along the surface. Generally, movement is accomplished through the plastic deformation of internal ice crystals. At the surface, expansion and travel is

accommodated by fracture. Glacial movement is ultimately driven by gravity (see question #2).

6. How fast do glaciers normally move? How fast can they move during a surge?
   Glaciers normally move from 10 to 300 m per year; they can surge as fast as 100 m per day.

7. Explain how the balance between ablation and accumulation determines whether a glacier advances or retreats.
   If accumulation of new snow to feed the glacier exceeds ablation (loss of ice due to melting or sublimation), the glacier will grow and advance; if the reverse is true, the glacier will "retreat" (disappear).

8. How can a glacier continue to flow downhill even though its toe is retreating?
   Downhill flow of ice is driven by gravity and is not interrupted by melting that occurs elsewhere. The toe is not truly retreating uphill but is merely melting or sublimating.

9. How does a glacier transform a V-shaped river valley into a U-shaped valley?
   Glaciers and their abrasive sedimentary load not only carve into the floor of a valley but also hollow out the sides.

10. Discuss how hanging valleys can develop and why they cannot be explained by the concept of a graded stream.
    Valley glaciers carving through a tributary to a major stream will not cut as deeply as the floor that is cut by the glacier in the main stream valley. Flowing water will not sustain hanging valleys, because the jump in slope increases the velocity of the tributary stream, increasing the stream's capacity to erode and carry sediment.

11. Describe the various kinds of glacial deposits. Be sure to note the materials from which the deposits are made and the landforms that result from deposition.
    Glacial materials include erratic boulders and unsorted tills dropped directly from the glacier, in addition to outwash sands and gravels from meltwater streams, glacial lake sediments, and windborne loess. Landforms include end moraines, dropped ridges of till at the frontal margin of a glacier that form when a glacier ceases to advance and melts away; lodgment till, the smeared remains of an end moraine steamrolled by an advancing glacier; and ground moraine, a plain of till released over the broad area of glacial recession. Drumlins are asymmetric hills of till shaped by glacial ice (with a steeper slope in the upstream direction).
    Kettles are depressions formed when till buries ice, which later melts and flows outward. Meltwater streams deposit outwash consisting of stratified sediments, and varved sediments may accumulate at the bottom of meltwater-fed lakes. Eskers are sorted sediments that reflect the course of a stream that existed on the bottom surface of a glacier.

12. How do the crust and mantle respond to the weight of glacial ice? What happens to the crust when a glacier melts away?

   Thick glacial ice may cause local subsidence of the lithosphere, which presses down into the softer asthenosphere below. When the glacier melts, the lithosphere rebounds to its prior elevation, and the asthenosphere flows in underneath the rising lithosphere.

13. How was the world different during the glacial advances of the Pleistocene ice age? Be sure to mention the relation between glaciations and sea level.

   Earth was much colder, and much of northern North America was covered in glacial ice; in the Northern Hemisphere, climatic zones shifted southward. Just to the south of the glaciers, conditions were wetter than normal, but the tropics were unusually dry. Because so much water was bound into glacial ice, sea level was considerably lower than today.

14. How was the standard four-stage chronology of U.S. glaciations developed? Why was it so incomplete? How was it modified with the study of marine sediment?

   There were four distinct layers of till, of differing ages, separated by soils with fossil remains suggesting returns to warmer climate. The record of advances and retreats, like so much of the terrestrial sedimentary record, is vastly incomplete because of erosion of many glacial deposits. Marine sediments can be tested for oxygen isotope ratios that serve as gauges of past ocean temperatures; these sediments also contain distinctive fossil assemblages associated with warm- or cool-water settings.

15. Were there ice ages before the Pleistocene? If so, when?

   There were ice ages in the late Paleozoic, late Proterozoic, and early Proterozoic, to name three examples.

16. What are some of the long-term causes that lead to ice ages?

   The proportion of carbon dioxide (a greenhouse gas) and plate tectonic activity are the major factors affecting long-term climate. Active mid-ocean ridge volcanism can increase the greenhouse effect by adding carbon dioxide to the atmosphere. Similarly, rapid ridge volcanism increases the volume of the ridges and can push sea level over the continents. The configuration of continents is also important; large continental masses near the poles favor the possibility of terrestrial glaciers.

20. Describe the three kinds of Milankovitch cycles. How does each affect the amount of sunlight received on Earth?

   Variation in the eccentricity of Earth's orbit affects the evenness of global insolation through the year. Additionally, Earth travels more slowly when it is more distant from the Sun, spending proportionately more time near aphelion in a more eccentric (elliptical) orbit.

   Variation in tilt affects seasonality. A small tilt means the poles will not receive a warm summer and favors glacial advance; however, no tilt would mean a mild winter, so it would be difficult for ice to accumulate.

Precession affects the timing of equinoxes and solstices. Summer will be cooler in the Northern Hemisphere if it occurs in conjunction with aphelion.

Test bank

1. Ice is a substance with a high albedo, which means it _____.
    A. requires much heat to raise its temperature by 1° C
    B. absorbs most of the light that falls upon it
    C. reflects most of the light that falls upon it
    D. strongly refracts the light that falls upon it

2. Because glacial advance is driven by gravity, it is impossible for glaciers to advance over perfectly flat terrain.
    A. true   B. false

3. Cirques and horns are features associated with _____.
    A. mountain glaciation   B. continental glaciation
    C. glacial outwash deposits D. loess deposits

4. Fine silt and clay are the characteristic sediment sizes of _____.
    A. mountain glaciation   B. continental glaciation
    C. glacial outwash deposits D. loess deposits

5. Eskers and drumlins are features associated with _____.
    A. mountain glaciation   B. continental glaciation
    C. glacial outwash deposits D. loess deposits

6. Stratified, sorted sand and gravel are deposited by _____.
    A. mountain glaciers   B. continental glaciers
    C. glacial outwash streams D. wind

7. Glacial ice exhibits _____ behavior near the top, but _____ behavior beneath a depth of 60 m.
    A. brittle; ductile
    B. ductile; brittle
    C. solid; liquid
    D. plastic; elastic

8. A glacier will always advance from its source area if the rate of accumulation is greater than the rate of _____.
    A. subsidence   B. melting
    C. ablation   D. abrution

9. The first scientist to theorize the past presence of glaciers in Europe was _____.
    A. Isaac Newton   B. Charles Darwin
    C. Walter Alvarez   D. Louis Agassiz

10. At the present, glaciers cover about _____ of the surface of the continents.
    A. 1%   B. 5%   C. 10%   D. 20%

11. The current interglacial interval began a little more than _____ years ago.
    A. 1,000   B. 10,000   C. 100,000   D. 1,000,000

12. Which statement about glacial ice is correct?
    A. It is a mineral.
    B. It is not a mineral because it does not have a fixed crystalline structure.
    C. It is not a mineral because it is an organic substance.
    D. It is not a mineral because its chemical composition is too variable.

13. An intermediate product in the transformation of snow to glacial ice is _____.
    A. firn   B. sublimation   C. ablation   D. terminus

14. All glaciers move by sliding along their bottom surface.
    A. true   B. false

15. A bowl-shaped depression formed by a mountain glacier is termed a(n)_____.
    A. arête   B. cirque   C. horn   D. tarn

16. A lake that forms within a bowl-shaped depression formed by a mountain glacier is termed a(n)_____.
    A. arête   B. cirque   C. horn   D. tarn

17. A remnant ridge separating two bowl-shaped depressions formed by a mountain glacier is termed a(n)_____.
    A. arête   B. cirque   C. horn   D. tarn

18. An angular peak surrounded by three or more bowl-shaped depressions formed by a mountain glacier is termed a(n)_____.
    A. arête   B. cirque   C. horn   D. tarn

19. Valleys carved by glaciers tend to be shaped like the letter _____, whereas valleys carved by water tend to shaped like the letter _____.
    A. "V"; "U"
    B. "V"; "C"
    C. "U"; "V"
    D. "V"; "I"

20. A hanging valley is formed when a _____.
    A. smaller glacially carved valley intersects a larger glacially carved valley
    B. smaller stream-cut valley intersects a larger stream-cut valley
    C. smaller stream-cut valley intersects a larger glacially carved valley
    D. stream-cut valley is on the upthrust side of a normal fault

21. When sea level rises, causing the ocean to fill a glacially carved valley, a _____ results.
    A. estuary   B. tarn   C. fjord   D. ford

22. Sediments deposited directly by glaciers as they melt are characterized by _____.
    A. uniformly coarse grain size   B. uniformly fine grain size
    C. very poor sorting   D. graded bedding

23. Sediments deposited directly by glaciers as they melt are termed _____.
    A. firn   B. loess   C. outwash   D. till

24. Sediments deposited by meltwater streams form stratified deposits termed _____.
    A. firn   B. loess   C. outwash   D. till

25. Wind blows finer particles long distances from glacial environments, where they settle out to form _____.
    A. firn   B. loess   C. outwash   D. till

26. Rapid volcanism along mid-ocean ridges _____.
    A. favors extensive continental glaciation
    B. reduces the likelihood of extensive continental glaciation
    C. cools the Earth because of the massive amount of volcanic ash extruded along the ridges, but not substantially enough to allow ice sheets to expand

27. An important long-term factor determining whether or not glacial ice will form on the continents has likely been the proportion of which gas in the atmosphere?
    A. oxygen   B. nitrogen
    C. carbon dioxide   D. carbon monoxide

28. The effect that periodic changes in Earth's orbital eccentricity and in the magnitude and direction (precession) of Earth's axial tilt have on the advance and retreat of ice sheets was first proposed by _____.
    A. Richter   B. Milankovitch
    C. Mohorivic   D. Lyell

29. By increasing the albedo of the Earth, global ice sheets produce conditions that are _____ to their further advance, thus providing an example of _____ feedback.
    A. detrimental; positive
    B. detrimental; negative
    C. favorable; positive
    D. favorable; negative

30. Areas on the southern margins of the continental glaciers of the Northern Hemisphere were much _____ during Plio-Pleistocene glaciations than they are today, as suggested by evidence of large _____ during this time.
    A. warmer; tropical rainforests
    B. wetter; tropical rainforests
    C. drier; deserts
    D. wetter; pluvial lakes

Chapter 23
Global Change in the Earth System

Learning objectives

1. Students should have an appreciation of the complexity and degree of interconnectedness of physical and biological systems on Earth. Earth has been changing since its beginnings, and students should be able to contrast the first crust and first atmosphere with those of the modern world.

2. Some changes during Earth history have been unidirectional (differentiation of crust, mantle, and core, and oxygenation of the atmosphere), whereas many other systems fluctuate or cycle from one state to another (such as global climate, which has experience a number of periods of greenhouse and icehouse conditions).

3. The rock cycle is an important theme in geology. Any of the three major rock types may be recycled to form rocks belonging to the other two major types (as well as a new lithology of the same type, such as phyllite being metamorphosed into schist). Other important physical systems in which change and fluctuation have been historically studied include the water cycle, global sea level, global climate, topographic relief (a balance of uplift and erosion), and the formation of supercontinents by the actions of plate tectonics.

4. The carbon cycle is an important example of a biogeochemical cycle. Carbon passes back and forth among the biosphere, atmosphere, hydrosphere, and lithosphere. Carbon dioxide is an important greenhouse gas, helping to keep the Earth warm through absorption of infrared radiation to space. Had liquid water not formed at the surface, most carbon on Earth would have persisted as carbon dioxide in the atmosphere (where it originally built up due to volcanic outgassing). This would have made conditions too hot for life as we know it.

5. Human industrial activity has led to the destruction of terrestrial ecosystems, the development of atmospheric smog over cities, and a hole in the ozone layer. Human emissions of greenhouse gases may lead to future global warming.

Summary from the text

We refer to the global interconnecting web of physical and biological phenomena on Earth as the Earth system. Because the asthenosphere is warm and soft enough to flow and because the surface of the Earth is warmed enough by the Sun to maintain water in liquid form, the Earth system is dynamic.

Global change refers to the transformations or modifications of physical and biological components of the Earth system through time. Unidirectional change results in transformations that never repeat, while cyclic change involves repetition of the same steps over and over.

Examples of unidirectional change include the gradual evolution of the solid Earth from a homogeneous collection of planetesimals to a layered planet on which lithospheric plates move; the formation of the oceans and the gradual change in the composition of the atmosphere; and the evolution of life.

Examples of physical cycles that take place on Earth include the supercontinent cycle, during which continents come together and later break apart; the sea-level cycle, during which the sea level rises and falls; the rock cycle, during which atoms travel among different rock types; and the landscape cycle, during which internal processes cause uplift and external processes cause erosion.

A biogeochemical cycle is the passage of a chemical among nonliving and living reservoirs. Global change occurs when something changes the relative proportions of the chemical in different reservoirs.

During the hydrologic cycle, water cycles between the oceans, the atmosphere, glaciers, surface and groundwater, and living organisms. If the climate cools, more water is stored in glacial reservoirs and the sea level falls, while if the climate warms, more water is stored in the oceans and the sea level rises.

During the carbon cycle, carbon cycles between the oceans, the atmosphere, living organisms, fossil fuels, and limestone. Burning fossil fuels returns carbon in the form of carbon dioxide gas to the atmosphere. The chemical weathering of rock absorbs carbon dioxide and removes it from the atmosphere. Atmospheric carbon dioxide can also dissolve in the ocean, where it may precipitate as calcite and become limestone. Carbon dioxide is a greenhouse gas, so changes in atmospheric concentration may affect the climate.

Tools for documenting global climate change, the transformations or modifications of Earth's climate through time, include the stratigraphic record, paleontology, oxygen isotope ratios, bubbles in ice, growth rings, and human history.

Studies of long-term climate change (measured in millions of years) show that at times in the past the Earth experienced greenhouse (warmer) periods, while at other times there were icehouse (cooler) periods. Factors leading to long-term climate change include the positions of the continents, volcanic activity, the uplift of land (which controls rates of chemical weathering), and the formation of materials (such as coal and limestone) that remove carbon dioxide. Negative feedback appears to prevent a runaway greenhouse effect on Earth.

Short-term climate change can be seen in the record of the last million years. In fact, during only the past 15,000 years, we see that the climate has warmed and cooled a few times. Causes of short-term climate change include fluctuations in solar radiation, changes in Earth's orbit and tilt, changes in reflectivity, and changes in ocean currents.

Mass extinction, a catastrophic change in biodiversity, may be caused by bolide impact or by intense volcanic activity associated with a superplume. The impact of a large bolide at the site of the Yucatan peninsula may have led to the extinction of the dinosaurs at the end of the Cretaceous period.

During the last two centuries, humans have changed landscapes, destroyed ecosystems (such as rainforests), and added pollutants to the land, air, and water at rates faster than the Earth system can process and neutralize. These changes result in a variety of societal problems, such as smog, water contamination, acid rain and runoff, more radioactive materials, and ozone depletion.

The addition of carbon dioxide and methane to the atmosphere may be causing global warming, which could shift climate belts, cause stronger storms, lead to a rise in sea level, and shut off thermohaline oceanic currents.

In the future, in addition to climate change (perhaps leading to another ice age?), the Earth will witness a continued rearrangement of continents resulting from plate tectonics and will likely suffer the impact of bolides. The end of the Earth may come when the Sun becomes a red giant star; the Earth will then vaporize, perhaps to become part of another solar system in the future.

Answers to review questions

1. Why do we use the term "Earth system" to describe the processes operating on this planet?

    Many biotic and abiotic processes are complexly interrelated.

2. How have the Earth's crust and atmosphere changed since they first formed?

    Early in its history, the Earth was hot enough to melt iron. Molten iron migrated toward the center of the Earth, differentiating the core from the mantle-like exterior. The first crust was a thin skin which was subjected to subduction and remelting. As the Earth cooled, subducted material from the surface no longer completely melted, and partial melting produced magmas that were more silicic than the ultramafic chemistry of the initial mantle and crust. Basaltic magmas formed oceanic crust, and intermediate and silicic magmas crystallized to form buoyant continental crust.

    Earth's primordial atmosphere of hydrogen and helium was likely lost to space; a secondary atmosphere dominated by the volcanic gases water vapor, carbon dioxide, sulfur dioxide, and nitrogen lost the first three of these to the oceans which formed as the Earth cooled. Oxygen was added by biotic photosynthesis to the remnant nitrogen, and these two gases dominate the modern atmosphere.

3. What processes control the rise and fall of sea level on Earth?

    Global sea level is primarily controlled by the volume of glacial ice on continents, which has a negative effect, and the volume of mid-ocean ridge volcanoes, which has a positive effect. At times of rapidly moving plates (abundant mid-ocean ridge volcanism), excess carbon dioxide leads to a stronger greenhouse effect, warming the ocean and melting glaciers (if any are present), which raises sea level by adding ocean water. Rapid volcanism also produces thick mid-ocean-ridge volcanic chains, which displace water onto the continents. Conversely, when sea-floor spreading rates are low, ridge volume is small and the atmospheric level of carbon dioxide is also low. The resultant reduction in the greenhouse effect favors the formation of continental glaciers, freezing out water that is now unavailable to the ocean.

4. How does carbon cycle through the various Earth systems?

    Carbon dioxide in the atmosphere dissolves in the ocean to form carbonate (and bicarbonate) ions. These ions are removed by a variety of organisms to produce calcite and aragonite skeletons, which collect as fragments and grains at the bottom after the organisms die, eventually forming limestone. Carbon dioxide can also be removed, either directly from the atmosphere or from solution in the ocean, through biogenic photosynthesis, to produce organic carbon. Organic carbon may become incorporated into the rock record in shale, oil, and coal, but some may be released to the environment through animal respiration and flatulence. Weathering of silicate rocks removes atmospheric carbon dioxide, producing bicarbonate ions. Burning fossil fuels releases atmospheric carbon dioxide, as do volcanic eruptions.

5. How do paleoclimatologists study ancient climate change?

The record of sedimentary rocks can be used to decipher ancient climate change, because certain rocks are characteristic products of specific environments; further, fossils provide environmental clues because many organisms have narrow environmental tolerances. Oxygen isotope ratios in ice and carbonate sediments provide a proxy for average temperatures. Ancient air bubbles may reveal atmospheric carbon dioxide levels (and thus the effectiveness of Earth's greenhouse). Variations in growth rings and recorded human history can be used to infer climate change in the very recent past.

6. Contrast icehouse and greenhouse conditions.

Greenhouse climate is warmer than today's (especially at the poles) and is characterized by a high atmospheric concentration of carbon dioxide and relatively high sea level, with no continental glaciers at the poles. In contrast, icehouse conditions are colder, with permanent ice present at the poles and relatively low levels of atmospheric carbon dioxide and low sea level.

7. What are the possible causes of long-term climatic change?

1. The sizes and positions of the continents are important; in order for large continental glaciers to grow it is favorable to have large continents in regions near the poles; small continents bathed in tropical oceans favor greenhouse conditions.

2. Volcanic activity: volcanoes emit carbon dioxide, which adds to the greenhouse effect.

3. Orogenic activity: uplifted areas are sites of intense weathering, and chemical weathering draws down atmospheric carbon dioxide.

4. Production and burial of carbonate and organic carbon: limestone, coal, organic-rich shale, and oil contain carbon, so when produced in vast quantities and buried, they keep carbon from reaching the atmosphere as carbon dioxide. Carbon burial is a check on the greenhouse effect.

8. What climatic changes may have been responsible for the birth and growth of the first civilization in Europe and Asia?

The timings of many early flourishing civilizations are in accord with the Holocene climatic optimum of 5000 to 6000 years ago, when climate was warmer and wetter than today.

9. What four factors explain short-term climatic changes?

1. The abundance of sunspots (cool spots on the surface of the Sun, which may represent magnetic storms) varies over the course of a decade or so, and may affect total incident solar radiation.

2. The Milankovitch cycles: Earth's orbital shape, magnitude of tilt, and direction of tilt vary over cycles with periods in the tens of thousands of years.

These parameters influence whether glaciers are likely to descend over the continents or melt.

   3. Earth's albedo can be increased by an increase in aerosols (such as volcanic ash), cloud cover, surface ice, or the spread of deserts and grasslands over land that was once forested.

   4. Ocean currents may change course, altering the hydrologic system that brings warmth to some areas, and cold, dry conditions to others.

10. Give some examples of events that cause catastrophic change.
   Bolide impact, mass extinctions, explosive or hyperactive volcanism, sudden episodes of global warming or cooling.

11. Give some examples of how humans have changed the Earth.
   Extraction of rock and groundwater, overhunting and overfishing, destruction of forests and grasslands, and pollution of the air, streams, and oceans have led to subsidence, increased mass wasting, famine, high rates of biotic extinction, acid rain, smog, a hole in the ozone layer, etc.

12. What is the ozone hole, and how does it affect us?
   The ozone hole is an large opening in the stratospheric ozone layer over Antarctica (a smaller hole sits atop the Arctic) caused by the reaction of ozone with anthropogenic chlorofluorocarbons.

13. Describe how carbon dioxide–induced global warming takes place, and how humans may be responsible. What effects might global warming have on the Earth's system?
   Carbon dioxide is an important greenhouse gas, trapping infrared radiation emitted from the Earth. Human burning of fossil fuels releases carbon dioxide into the atmosphere. Global warming can raise sea level (by melting glaciers) and produce coastal inundation.

14. What are some of the likely scenarios for the long-term future of the Earth?
   Unless destroyed by impact, Earth will be consumed by the red giant stage of our Sun's evolution, approximately 5 billion years from now.

Test bank

1. Nearly all biological and chemical systems on Earth are disconnected and are best studied in mutual isolation.
   A. true   B. false

2. The presence of liquid water on Earth's surface _____.
   A. is unremarkable; many other bodies in the solar system have liquid water
   B. is unique within the solar system, but is an inherent feature of Earth that would be found regardless of the Earth-Sun distance
   C. is unique within the solar system; liquid water would not have developed had the Earth been much closer or farther from the Sun

3. All changes within the Earth system are unidirectional and cannot be reversed.
    A. true   B. false

4. Cyclical global change includes the formation of core, mantle, and crust from a homogeneous mixture, a state to which one day Earth will return.
    A. true   B. false

5. Which of the following types of global change is not reversible?
    A. orogenic uplift
    B. melting and crystallization of sedimentary rock to form igneous rock
    C. evolution of life on Earth
    D. flooding of the continents due to global warming

6. Uplifted areas are subjected to erosion at the surface; this provides an example of _____.
    A. positive feedback
    B. negative feedback
    C. neutral feedback

7. Which of the following processes releases carbon dioxide to the atmosphere?
    A. photosynthesis   B. weathering of silicate rocks
    C. volcanism   D. deep burial of peat to produce coal

8. Which of the following processes removes carbon dioxide from the atmosphere?
    A. exhalation by animals   B. burning of coal and other fossil fuels
    C. volcanism   D. photosynthesis

9. Which of the following processes may be responsible for short-term cooling, yet in the long term favors warm, greenhouse climate?
    A. atmospheric carbon dioxide
    B. extensive ice sheets
    C. volcanism
    D. large continents located near the poles

10. Pollen has changed over time through evolution, but the pollen of individual species cannot be used to interpret ancient environments because most species inhabit all environments.
    A. true   B. false

11. An increase through time in the proportion of heavy oxygen (O-18) in a sequence of carbonate sediments implies _____.
    A. the sediments are young enough that they were exposed to radioactive testing
    B. the organisms that secreted the carbonate favored O-18 over the far more abundant O-16
    C. Earth was becoming warmer over time
    D. Earth was becoming cooler over time

12. There is a correlation in Earth history between warm periods and _____.
    A. relatively large amounts of carbon dioxide in the atmosphere
    B. low sea level
    C. an abundance of heavy oxygen isotopes in sediments deposited at the time
    D. widespread formation of coal

13. During most of the Mesozoic, the Earth was _____.
    A. about as warm as today
    B. covered in extensive continental glaciers
    C. about as warm as the Holocene climatic optimum
    D. significantly warmer than at any time within the last few million years

14. The region of greatest temperature contrast, as compared between greenhouse and icehouse periods, would be found _____.
    A. on the equator   B. at mid-latitudes   C. at the poles

15. The difference in average temperatures between the poles and the equator during most of the Mesozoic era was _____ that seen today.
    A. greater than   B. less than   C. equal to

16. The factors that affect Earth's long-term climate are different from those that affect climate on shorter time scales.
    A. true   B. false

17. The earliest large-scale civilizations arose during the _____.
    A. Younger Dryas cold interval
    B. Holocene climatic optimum
    C. medieval warm period
    D. little ice age

18. Which of the following is an important long-term (acting over spans of tens of millions of years) factor in global climate change?
    A. decrease in albedo due to soot from volcanic eruptions
    B. variation in Milankovitch orbital parameters
    C. variation in the abundance of sunspots
    D. atmospheric concentration of carbon dioxide

19. A crater discovered in the Yucatan peninsula in Mexico appears to be the right age to be a result of an impact which brought about the extinction of numerous species at the end of the _____ period.
    A. Silurian   B. Jurassic
    C. Cretaceous   D. Tertiary

20. The hole in the ozone layer has been brought about by anthropogenic emissions of _____.
    A. carbon dioxide   B. methane
    C. chlorofluorocarbons (CFCs)   D. polychlorinated biphenyls (PCBs)

Answers to Multiple-Choice Questions

Chapter 1

1. B
2. A
3. B
4. A
5. C
6. D
7. A
8. C
9. B
10. B
11. C
12. C
13. D
14. D
15. C
16. A
17. C
18. C
19. C
20. B
21. C
22. C
23. C
24. A
25. B
26. B
27. C
28. C
29. D
30. D

Chapter 2

1. B
2. B
3. C
4. C
5. B
6. A
7. C
8. B
9. B
10. B
11. C
12. A
13. A
14. C
15. B
16. A
17. C
18. B
19. D
20. D
21. B
22. A
23. B
24. B
25. D
26. A
27. D
28. C
29. A
30. D

Chapter 3

1. D
2. C
3. C
4. B
5. B
6. B
7. C
8. C
9. D
10. B
11. B
12. B
13. C
14. A
15. C
16. D
17. D
18. A
19. B
20. A

Chapter 4

1. D
2. C
3. A
4. C
5. A
6. A
7. D
8. A
9. C
10. B
11. B
12. C
13. C
14. B
15. B
16. B
17. A
18. C
19. B
20. A
21. B
22. C
23. C
24. B
25. A
26. C
27. D
28. C
29. B
30. C
31. C
32. B
33. B
34. B
35. B
36. D
37. A
38. C
39. B
40. C
41. C
42. C
43. A
44. B
45. B
46. A
47. B
48. C
49. B
50. B

Chapter 5

1. B
2. D
3. C
4. A
5. B
6. C
7. C
8. B
9. D
10. B
11. B
12. B
13. B
14. B
15. B
16. C
17. C
18. C
19. B
20. A
21. C
22. B
23. B
24. D
25. B
26. B
27. B
28. C
29. C
30. A

Chapter 6

1. C
2. B
3. D
4. A
5. C
6. C
7. D
8. A
9. C
10. D
11. A
12. B
13. D
14. C
15. B
16. C
17. B
18. A
19. A
20. A
21. C
22. C
23. D
24. D
25. C
26. B
27. D
28. C
29. B
30. C
31. A
32. B
33. D
34. A
35. B
36. B
37. B
38. A
39. C
40. A

Chapter 7

1. C
2. B
3. C
4. C
5. C
6. B
7. D
8. C
9. C
10. B
11. A
12. D
13. B
14. A
15. C
16. D
17. B
18. D
19. C
20. D
21. C
22. B
23. C
24. A
25. B
26. A
27. C
28. B
29. A
30. C
31. D
32. D
33. C
34. D
35. B
36. A
37. A
38. B
39. B
40. C
41. A
42. D
43. B
44. D
45. C
46. D
47. B
48. A
49. C
50. B

Chapter 8

1. B
2. B
3. A
4. C
5. D
6. C
7. D
8. C
9. A
10. B
11. D
12. C
13. C
14. B
15. B
16. C

17. C
18. B
19. B
20. C
21. D
22. B
23. C
24. D
25. B
26. D
27. D
28. C
29. C
30. B
31. D
32. C
33. C
34. A
35. A
36. D
37. D
38. A
39. C
40. C

Chapter 9

1. C
2. C
3. A
4. A
5. D
6. B
7. D
8. A
9. A
10. B
11. B
12. A
13. D
14. B
15. A
16. A
17. D
18. C
19. C
20. C
21. B
22. A

23. A
24. A
25. C
26. B
27. B
28. B
29. D
30. C
31. B
32. A
33. D
34. D
35. A
36. C
37. B
38. A
39. B
40. B

Chapter 10

1. A
2. B
3. B
4. A
5. B
6. D
7. A
8. C
9. C
10. B
11. C
12. C
13. C
14. A
15. B
16. B
17. C
18. C
19. D
20. A
21. D
22. D
23. B
24. A
25. B
26. C
27. D
28. B

29. A
30. B
31. C
32. B
33. A
34. C
35. A
36. B
37. D
38. A
39. C
40. B

Chapter 11

1. B
2. A
3. C
4. B
5. D
6. C
7. B
8. A
9. B
10. B
11. A
12. A
13. A
14. B
15. B
16. A
17. B
18. B
19. A
20. D
21. C
22. C
23. C
24. D
25. C
26. A
27. D
28. C
29. B
30. B
31. B
32. B
33. D
34. B

35. C
36. C
37. A
38. C
39. B
40. A

Chapter 12

1. D
2. B
3. C
4. C
5. B
6. B
7. D
8. D
9. B
10. B
11. C
12. D
13. B
14. D
15. A
16. A
17. B
18. C
19. B
20. B
21. C
22. C
23. C
24. A
25. B
26. B
27. B
28. C
29. D
30. C
31. D
32. A
33. D
34. C
35. D
36. A
37. C
38. B
39. C
40. B

Chapter 13

1. B
2. B
3. A
4. D
5. C
6. D
7. C
8. D
9. C
10. B
11. D
12. D
13. A
14. B
15. A
16. B
17. C
18. D
19. A
20. C
21. B
22. B
23. D
24. D
25. D
26. A
27. B
28. D
29. C
30. D
31. C
32. C
33. B
34. C
35. C
36. D
37. D
38. C
39. D
40. A
41. B
42. A
43. C
44. D
45. C
46. B
47. A

48. B
49. C
50. C

Chapter 14

1. B
2. A
3. D
4. D
5. D
6. B
7. C
8. A
9. B
10. B
11. C
12. A
13. B
14. C
15. C
16. B
17. C
18. A
19. B
20. C
21. A
22. B
23. D
24. C
25. A
26. B
27. C
28. C
29. C
30. B
31. C
32. A
33. B
34. C
35. B
36. C
37. B
38. B
39. C
40. B

Chapter 15

1. B
2. B
3. C
4. A
5. C
6. B
7. A
8. B
9. B
10. C
11. B
12. C
13. B
14. D
15. A
16. A
17. C
18. B
19. D
20. A
21. C
22. D
23. C
24. D
25. D
26. B
27. B
28. D
29. B
30. B

Chapter 16

1. B
2. D
3. A
4. C
5. D
6. B
7. B
8. D
9. C
10. A
11. B
12. B
13. C
14. C
15. B
16. A
17. D
18. D
19. C
20. D

Chapter 17

1. B
2. A
3. D
4. C
5. B
6. A
7. C
8. D
9. B
10. B
11. A
12. C
13. B
14. C
15. D
16. A
17. A
18. B
19. C
20. B
21. B
22. D
23. B
24. B
25. D
26. B
27. B
28. B
29. A
30. B
31. C
32. B
33. D
34. C
35. A
36. B
37. C
38. B
39. A
40. C

Chapter 18

1. C
2. B
3. C
4. B
5. B
6. A
7. B
8. C
9. D
10. A
11. C
12. D
13. C
14. B
15. B
16. C
17. B
18. B
19. C
20. C
21. D
22. C
23. D
24. B
25. A
26. C
27. B
28. B
29. C
30. B
31. D
32. D
33. C
34. A
35. B
36. C
37. C
38. C
39. B
40. C

Chapter 19

1. C
2. C
3. B
4. A

5. C
6. D
7. B
8. C
9. C
10. D
11. B
12. A
13. B
14. D
15. C
16. C
17. B
18. A
19. C
20. B
21. D
22. C
23. B
24. A
25. B
26. C
27. A
28. B
29. C
30. C
31. A
32. B
33. A
34. C
35. C
36. B
37. D
38. B
39. B
40. B

Chapter 20

1. B
2. D
3. A
4. C
5. B
6. C
7. B
8. B
9. A
10. D

11. D
12. B
13. A
14. C
15. C
16. B
17. D
18. C
19. B
20. C
21. B
22. B
23. D
24. A
25. B
26. C
27. B
28. A
29. B
30. B
31. A
32. C
33. B
34. A
35. C
36. A
37. D
38. C
39. A
40. B

Chapter 21

1. B
2. C
3. D
4. B
5. B
6. B
7. A
8. C
9. B
10. B
11. B
12. A
13. D
14. C
15. B
16. B

17. A
18. D
19. B
20. C
21. B
22. C
23. B
24. B
25. A
26. D
27. A
28. C
29. D
30. C

Chapter 22

1. C
2. B
3. A
4. D
5. B
6. C
7. A
8. C
9. B
10. C
11. B
12. A
13. A
14. B
15. B
16. D
17. A
18. C
19. C
20. A
21. C
22. C
23. D
24. C
25. B
26. B
27. C
28. B
29. D
30. B

Chapter 23

1. B
2. C
3. B
4. B
5. C
6. B
7. C
8. D
9. C
10. B
11. D
12. A
13. D
14. C
15. B
16. A
17. B
18. D
19. C
20. C